ROAD RESEARCH

ntrol
onditions

PARED
EARCH GROUP

981

ERATION AND DEVELOPMENT

The Organisation for Economic Co-operation and Development (OECD) was set up under a Convention signed in Paris on 14th December 1960, which provides that the OECD shall promote policies designed:

— to achieve the highest sustainable economic growth and employment and a rising standard of living in Member countries, while maintaining financial stability, and thus to contribute to the development of the world economy;
— to contribute to sound economic expansion in Member as well as non-member countries in the process of economic development;
— to contribute to the expansion of world trade on a multilateral, non-discriminatory basis in accordance with international obligations.

The Members of OECD are Australia, Austria, Belgium, Canada, Denmark, Finland, France, the Federal Republic of Germany, Greece, Iceland, Ireland, Italy, Japan, Luxembourg, the Netherlands, New Zealand, Norway, Portugal, Spain, Sweden, Switzerland, Turkey, the United Kingdom and the United States.

Publié en français sous le titre:

RÉGULATION DE LA CIRCULATION
EN CONDITIONS DE SATURATION

***

FOREWORD

The Road Research Programme has two main fields of activity :

- promotion of international co-operation in the field of research on road transport, the co-ordination of research facilities available in Member countries and the scientific interpretation of the results of joint experiments ;
- International Road Research Documentation, a co-operative scheme for the systematic exchange of information on scientific literature and of current research programmes in Member countries.

The present programme is primarily concerned with defining the scientific and technological basis needed to assist governments of Member countries in decision-making on the most urgent road problems :

- development and evaluation of integrated urban and suburban transport strategies, taking into account economic, social, energy and environmental requirements ;
- traffic control systems, both in urban and rural areas, to optimise traffic operation and to enhance quality of service provided to road users ;
- formulation, planning and implementation of common overall strategies for safety ;
- planning, design and maintenance of the total road infrastructure, taking account of economic, social and technical developments and needs.

ABSTRACT

The report is the outcome of a study carried out by an OECD Road Research Group on Traffic Control in Saturated Conditions in which thirteen countries participated. It highlights the effects of traffic congestion in terms of energy wastage, environmental nuisance and driver annoyance. Firstly, the phenomena of saturation, congestion and over-saturation are defined together with the control objectives, parameters and measures which need to be considered. Subsequently, the present state-of-the-art is assessed, focussing on those concepts and control strategies which are promising. Thirdly, particular attention is devoted to future research needs. A set of Appendices provides the most recent information on traffic-actuated control models at isolated intersections, on-line models, off-line programmes, mathematical approaches and traffic corridor strategies. This international study shows how, with the help of modern computerised tools, the systematic application of traffic control strategies can improve the efficiency of existing road infrastructure and enhance its use. A new research approach is needed to improve the use of preventative and remedial control measures for saturated conditions within the framework of a balanced traffic and transportation policy.

PREFACE

Recent surveys in OECD countries suggest that traffic conditions at periods of peak demand have generally worsened. Reduced funds for programmes to alleviate the capacity limitations of existing road infrastructure in urban areas and along major long-distance corridors, as well as sustained traffic growth, are amongst the major factors contributing to recurring conditions of overloading. The scale and severity of these problems, coupled with the requirement to conserve energy by reducing potential sources of petrol wastage, have made it necessary to systematically assess the various possible strategies for 'congestion management'. The present study addresses this issue as part of an ongoing cooperative effort within the OECD Road Research Programme to advance the state-of-the-art in the field of traffic control systems.

Although congestion control has been a perennial task for generations of traffic engineers, few studies have been carried out to develop control strategies that are specifically adapted to handle conditions of saturation or over-saturation. It is only during the last few years that this control problem has gained prominence in international research. However, there are no ready-made solutions. Indeed, the traffic information needed, as well as the type of strategies, differ too much from those for non-saturated situations so as to confine development efforts to simple adaptations of conventional traffic control modes.

The Group began its investigation by analysing the processes which lead to over-saturation and identifying the control objectives to be set for this control case, as well as the measures of effectiveness. A distinction between preventative and curative measures is made and a hierarchy of strategies for the various situations (intersections, arterials, networks and corridors) in both urban and rural areas is developed.

In assessing the state-of-the-art proper, the study revealed some promising new (theoretical) avenues for developing control schemes for saturated traffic, but there is a lack of operational control modes. Most of the scientific work undertaken concerns the control of local intersections, and of motorways and/or traffic corridors. Information on congestion control of arterial roads and networks is scarce in regard to both offline and online control and the Group, therefore, attempted to identify empirical experience gained with pragmatically developed control schemes in Member countries. Special attention is devoted to the operation of public transport vehicles in congested conditions.

The report provides policy-makers and traffic engineers with an up-to-date assessment of traffic congestion management. The study also emphasizes the need for further research on strategy development and control models. In addition, work in this area should take into account driver responses, intervention thresholds and the response time of complex systems when dealing with online control. As more research results and operational experience become available, the assessment of traffic congestion problems and their potential solutions will be more reliable and effective.

TABLE OF CONTENTS

Page

I

INTRODUCTION

I.1 STATEMENT OF THE PROBLEM

For the control of traffic an array of strategies and algorithms is available, ranging from fixed time control at intersections to network co-ordination based on actual traffic flow data. These control modes, however, have all been designed to handle traffic under normal i.e. non-saturated conditions, the aim being to minimise delay and in some cases the number of stops. But OECD Member countries continue to be confronted with the perennial task of preventing or remedying recurrent conditions of traffic congestion at periods of peak demand. This problem calls for a different type of control.

Capacity restriction of the existing road infrastructure as well as the relative failure of broader actions to offset prevailing urbanisation and activity patterns and/or to promote public or semi-collective transport modes have tended to perpetuate the occurrence of oversaturated traffic conditions. Concomitantly, drivers have been individually and collectively subjected to increasingly longer delays and have experienced stop-and-go driving conditions in queues occurring not only on a steadily rising number of arteries in urban areas but also throughout whole networks and long-distance corridors. The driving public is now fully aware of the traffic management problems associated with oversaturated traffic conditions and the need to look for new solutions of minimising traffic congestion and its effects, especially its environmental aspects, is commonly accepted. Moreover, congestion management has become of prime importance as the requirement to conserve energy by reducing all potential sources of petrol wastage has come to the forefront. Consequently traffic scientists and engineers are called upon to develop measures that prevent (or delay the attainment of) saturation or remedy these situations as expeditiously as possible.

Congestion problems occur both on major arterials, especially those carrying commuter traffic, and in cities. Although they normally do not last much longer than one hour, their effect is enormous : congestion not only increases delay as demand exceeds capacity, it can also create a spin-off effect which is much more dangerous. Firstly, queues of waiting vehicles can block the network considerably. Secondly, as traffic breaks down the risks increase that an incident will further reduce capacity.

Table 1 shows these effects and how drastically only marginal increases in demand can lower the average speed in a city.

Table 1

EFFECT OF CONGESTION IN INNER LONDON

	1968-69	1971-72	1974-75	1977-78
Average speed, morning peak, Km/h	20.4	20.7	22.8	19.8
Daily traffic, vehicles	-	396,000	385,000	397,000

The reduction in traffic performance, in terms of cost per kilometre travelled using the formula cost/Km = 4.3 + 258/ v where v = travel speed in km/h and costs are given in pence (1979), is highlighted by the following difference between 74/75 and 77/78 : flow increase 3 %, speed decrease 9 %, cost per kilometre increase 12 %. Although the estimates are very rough, they give a good idea of the problems of congestion in cities.

A typical bottleneck on a motorway in the Netherlands causes a delay of 600 vehicle hours a day. Besides increased operating costs, work-hours lost etc. such bottlenecks cost about 75,000 litres of petrol per year.

Finally public transport vehicles deserve special attention. Not only the delay of a bus costs more than the delay of a car, but traffic management and public transport services are also affected by bus delays. For the Paris area, the following calculations of the effect of congestion have been made for 1977 :

- Costs of time lost by cars	£ 287,614,000
- Costs of time lost by bus passengers	£ 64,931,000
- Extra fuel costs for the cars	£ 136,000,000
- Extra fuel costs for the buses	£ 650,000
- Extra bus operating costs	£ 29,300,000

One of the problems in obtaining public support for research work on traffic control in oversaturated conditions is the under-estimation of the gravity and complexity of the problems which goes parallel with the belief that control-methods used for normal traffic conditions would automatically give solutions for congestion control. Indeed the nature of traffic control for saturated conditions may be essentially different from that of standard control modes. Under normal (non-saturated) conditions, traffic control is needed for reasons of safety, efficiency and prevention of congestion. Under saturated conditions, the throughput is essential. Therefore the objectives and strategies for each type of problem differ.

The last 20 years have seen an enormous expansion of traffic control strategies and equipment for unsaturated conditions which, of course, represented the design conditions sought by traffic engineers. Centralised and available hardware and software are dedicated to the task of minimising traffic delay under normal conditions. However, given the difference between both control situations it does not automatically follow that this infrastructure is also adequate for traffic control under saturated conditions.

Essentially, all control modes for local intersections try to give "green" only where there is traffic that can make an efficient use of it. In networks the strategies try to do this taking into account the knowledge about the arrival of traffic from an upstream intersection. However, all these strategies can be used under certain conditions only. One of these is that the network or the intersection is not congested. If this condition is not met the strategies fail, i.e. the original objectives cannot be met and the control mode used is not suitable to handle congestion.

As saturation is a situation where demand exceeds capacity it is necessary to know the demand distribution over the day. What is known about possible strategies shows that the greater the time horizon of predictive traffic information, the better the effectiveness of the strategy for saturated conditions. Consequently the equipment to be used for saturated control will be different from the standard control devices which collect little information about incoming vehicles held in queues. Indeed the strategies that have been developed for non-saturated conditions use much simpler optimisation procedures than those that promise good results for congestion control. Finally, comparing non-saturated with saturated conditions, a review of research and empirical results shows that the information on this fairly new topic is very scarce.

I.2 SCOPE AND OBJECTIVES OF THE STUDY

There is general agreement among Members of the Group that in many countries oversaturated conditions occur during peak periods, both in urban and rural areas, and that the standard control strategies are generally not designed to deal with such conditions. There is, however, a divergence of opinion on the parameters that should be used to optimise the traffic control modes for these conditions ; strategies are still in the experimental phase and the objectives to be attained have yet to be clearly defined.

The recognition of the need to give priority to public transport vehicles makes it all the more necessary to develop adequate strategies. It is evident that public transport vehicles are hampered by oversaturation but giving them priority can correspondingly lead to oversaturation.

The terms of reference of the Group were as follows :

"1. Define and classify oversaturated situations and their causes ; study the best parameters to describe these conditions, taking into account the demands of the various groups (pedestrians, children, public transport, etc.).

2. Make a survey of the various strategies used or under study.

3. Evaluate the results obtained in these studies with the aim of :

(a) defining the field data to be measured on line as an input for the application of the control modes ;

(b) suggesting measures on how to tackle the problem of oversaturation ;

(c) establishing a list of future research needs.

The study will examine traffic control on an areawide basis as well as for intersections and will cover both urban and rural situations."

This activity forms part of continuing interests and activities of the OECD Road Research Programme in the area of traffic control systems. Account has been taken of previous studies including (1) Area traffic control systems, (2) Electronic aids for freeway operation, (3) Traffic operation at sites of temporary obstruction, (4) Capacity of at-grade junctions, (5) Traffic corridor control, (6) Bus lanes and busway systems, (7) Integrated urban traffic management, (8) Traffic measurement methods. In addition this Group participated in the International Symposium on Traffic Control Systems held in Berkeley on 6th-9th August, 1979.

As regards item 3a the Group was not able to come to final conclusions. The reason is that the evaluation of the available information has shown that congestion control is still very much in the research phase.

The Group decided to focus on traffic control schemes for existing roads, motorways and networks and to exclude other methods to alleviate oversaturation such as urban planning, construction of new roads, overall traffic management and major traffic rerouting. It was agreed to consider the philosophy, strategies, modes, measures and parameters for control under saturated conditions disregarding infrastructural measures. Control for all major road elements ranging from intersections to complete networks would be considered, in both urban and rural (inter-city) situations. Account was taken of the different catagories of road-users, including public transport vehicles. Attention was given to the problem of defining and measuring "congestion", and consideration made of the use of queue length and position as a major indicator for describing oversaturated conditions.

The primary and secondary objectives of traffic control under saturated conditions, as well as the basic concepts involved, are given in Chapter II of this report. A hierarchy of strategy goals and operational objectives is also provided.

Chapter III provides a detailed review of existing strategies, summarising the state-of-art in this field. An attempt is made to describe recent work and new strategies in detail, while presenting only brief reviews of known and "traditional" concepts based on control modes for normal/non-saturated conditions. Theoretical and practical aspects of control of intersections, arterial roads, urban and inter-city networks are described. Special attention is paid to the specific public transport aspects.

Chapter IV contains a critical analysis presenting mainly a review of Chapter III. The shortcomings of the existing systems are discussed, and an emphasis given to the preventative measures with particular regard to critical intersections. The theoretical aspects are also treated in this chapter allowing a comparison of the theoretical models and field experience.

Chapter V attempts to specify improved strategies and/or study tools which are better suited to the problems of oversaturation. Recommendations on future research and development are made.

The nature of this report has led to a split-up in general chapters and more specific appendices aimed at those who are interested in more detailed information. The appendices deal with :

A : Terminology

B : Vehicle-and traffic-activated control models at isolated intersections.

C : On-line models

D : Off-line models (study tools)

E : Mathematical models

F : Traffic corridors

II

DEFINITION, CONTROL OBJECTIVES, PARAMETERS AND MEASURES

II.1 DEFINITION

The increasing volume of traffic on both urban and rural roads has put such pressure on existing facilities that one of their main functions, i.e. ensuring mobility, has been seriously impeded.

The very aim of traffic control is to provide the best traffic conditions for the users of the network. It is usual to rate these conditions in terms of "quality of service" and assess them on the basis of the results of a series of measurements. "Congestion", or more technically "saturation", is a state of these conditions. It is intuitively understood that congestion constitutes a poor quality of service and that it is associated with the existence of queues which increase over time, whether at a signalised intersection or at a bottleneck. The key point of interest in traffic control operation is clear : to identify when congestion starts, i.e. the saturation point, since the task of the traffic engineer is to avoid such a traffic state which is both below standard and inconvenient to road users. A good definition would, therefore, be as follows :

> "A network element is saturated when any new, even minor, deterioration in traffic conditions and/or increase in the traffic load introduce an abrupt drop in the quality of service of the network element and its associated control" (see Figure 1).

Saturation is, therefore, characterised by a discontinuity in the variation of the quality of service. This is what happens, for example, at a signalised junction at which vehicles have to stop more than once ; the mean waiting time then changes from part of a cycle to n times the cycle time. The same thing happens at a bottleneck which cannot handle all the traffic arriving at it : saturation then appears as a discontinuous flow of vehicles which have stop at least once before crossing the bottleneck. At the same time, an abrupt increase in travel time is observed, as soon as the vehicles form a queue.

In addition to the abrupt drop in the quality of service - a consequence of saturation - the phenomenon is amplified by its irreversibility ; this is shown in Figure 1. The lines of this figure show the place of the possible combinations of quality of service and demand. At any time the traffic situation can be represented by one point somewhere on these lines.

It will be observed that, as the demand increases, the quality of service decreases progressively from 1 to 2, where a break occurs (queues, blocking by turning traffic,..). It is at point 2 that the potential demand becomes greater than capacity. The fall 2-3 is controlled by two factors : firstly, by the time for which the demand persists and secondly, by the virtually instantaneous reduction in capacity ; even if the demand decreases beyond 3, the system is not reversible and the return path towards a saturated situation lies below the outward path. Saturation appears as a hysteresis curve, the surface of which represents the integral of the quality of service lost. The size of this surface is related to the drop in capacity and the time the phenomenon has persisted. Note that in time we can move upward and down either line 1-2 or 5-4. Lines 2-3 and 5-6 are of more theoretical value. No saturation represented by a point on 2-3 and 5-6 can last long. Note also that 2-3 and 5-6 are uni-directional in time. The fact of not having crossed 2 allows the system to operate on curve 6-2, instead of 4-5.

Figure 1 **TRAFFIC SATURATION POINT**

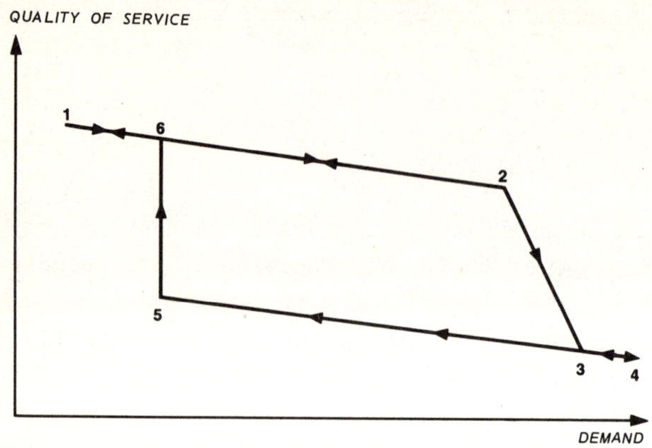

The graph used throughout this Chapter is designed to explain the relation between demand and quality of service as a function of both time and control strategy. The effectiveness of a control strategy can be expressed only when time is one of the parameters.

The performance or efficiency of a control strategy is the summation of the quality of service values measured at time intervals. This is related to the successive traffic situation as they change over time and which can be represented by a point moving along the lines depicted.

It should be noted that saturation usually occurs at particular places which are, effectively, the weakest points in the network. This can occur occasionally (car breakdown, delivery,..) or regularly. Generally, the reduction in capacity spreads very rapidly from this point, if adequate measures are not taken.

Since traffic flows along a system, made of interdependent elements, saturation may appear either in an isolated link of the network without affecting the whole system (localised saturation) or, because of interdependence, may be transmitted to the rest of the system, substantially disrupting its functional capacities (gangrenous saturation), without every affected link necessarily reaching its congestion threshold. NCHRP report 194 employs the terms "congestion, saturation and oversaturation" for what is termed here "saturation, local and gangrenous saturation", defining them on the basis of the queue formation mechanism, as shown in Table 2.

Saturation, or in other words, the abrupt drop in traffic characteristics, and its negative effects, must obviously be avoided (preventative treatment) or, if it occurs, a return to a normal situation must be sought (curative treatment). This problem should, therefore, be approached technically and investigated by means of a rational methodology to allow a full insight into its complex structure and, hopefully, provide a solution.

However, to talk about "treating saturation" seems a contradiction in terms, since the obvious aim of such a treatment would be precisely to avoid it by preventative measures ; it is thus more worthwhile to talk of "controlling the phenomenon", this control consisting in replacing the initial uncontrollable system by one giving better quality of service and a predictable, controllable and reversible mode of working.

Table 2

TRAFFIC DESCRIPTIONS

Unsaturated	Saturated		Gangrenous saturation
	Local saturation		
	Stable	Unstable	
No queue formation	Queue formation, but not growing Delay effects are local	Queue formation and growing, Delay effects still local (A transient state may be only of short duration)	Queue formation and growing to point where upstream intersection performance adversely affected

Note : An Annex defining terminology used in this text is added at the end of this report.

Preventative Measures

These control measures encompass two possible actions. First, increasing the capacity of the network and enhancing traffic control operation (point(2)) such that the capacity remains greater than demand.

The second possibility to be used when the first one is exhausted, consists of a trade-off between quality of service and demand, avoiding the abrupt fall in the former by means of traffic control. This strategy is represented by line 1-7 in Figure 2.

Figure 2 **PREVENTATIVE TREATMENT OF SATURATION**

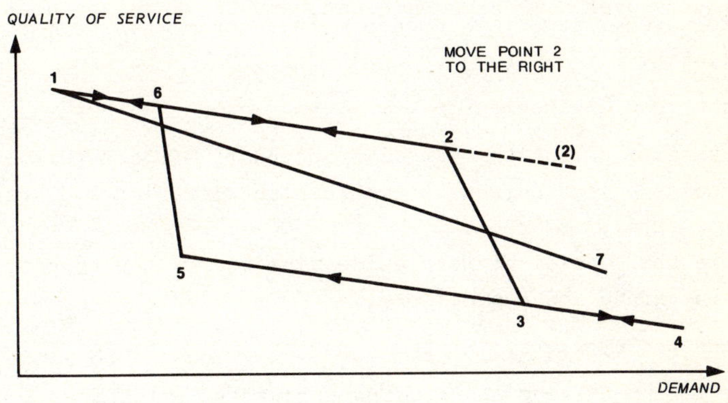

Curative (or palliative) Measures

The control in this case means - when the previous methods have been exhausted - moving the excess demand to zones where the congestion, artificially created and controlled, possesses minimum hysteresis cycles, i.e. to where capacity is least dependent on demand (see Figure 3).

Figure 3 **CURATIVE TREATMENT OF SATURATION**

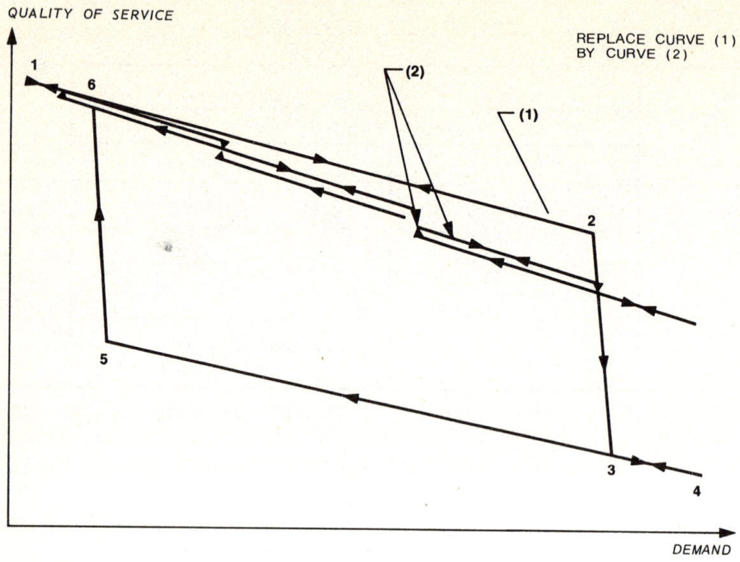

The disruption represented by the area 6-2-3-5 must be reduced at all costs, and traffic engineers should provide a system having a predictable mode of working.

Effects

A first step in the control effort is the identification of the conditions encompassed within the saturation denomination, which are mostly negative. The most evident negative effects are the following :

For the off-the-road environment :

a) excessive exhaust pollution, with its secondary effects (dirt, etc.) ;
b) excessive noise pollution, due to vehicle acceleration and braking, and over use of car horns, etc.,
c) reduction in the conditions for residential use of the street, with a consequent reduction in safety and convenience.

For pedestrians and cyclists :

a) pollution, making walking and cycling unpleasant activities ;
b) blocking of pedestrian crossings at intersections by queue overflows.

For the in-vehicle trip maker :

a) excessively low travel speeds - and hence longer travel times ;
b) average delays at intersections in excess of one cycle of the control system. This situation is often characterised by the appearance of residual queues at the intersection, the production of moving queues so restricted in their movement that their car-following instability is high, or the appearance of side-street blocking by these queues ;
c) excessive travel delays. Factors a) and b) compounded produce a situation which may be expressed as the occurrence of too many stops, namely more than one per link ;
d) excessive travel times for the transit user.

For the country as a whole :

a) an inefficient use of energy resources ;

b) inefficient movement of people and goods.

Objectives

An adequate road traffic management policy should aim to reduce these negative effects as much as possible. The goal pursued is to obtain a network flow pattern, which provides efficient accessibility and mobility at satisfactory levels of safety and comfort for all users, and which protects the residential function of each link.

A hierarchy of traffic objectives exists vis-à-vis these different goals, which may be attained by a coherent combination of policies and actions at different levels :

- The first level - urban planning - is the most aggregated, and encompasses decisions affecting the processes of regional planning, land-use, and transportation planning, all of which are linked with the location, size and type of the traffic generators.
- In the second hierarchical level - traffic planning or management - are included those objectives related to traffic organisation ; these comprise such measures, as one-way traffic roads, parking policies, street hierarchy, etc., as well as modal split objectives (with special reference to the split between public and private transport).
- Finally, in the third level - traffic engineering - may be included decisions relating to such aspects of traffic control as marking, signs, and signals, the objectives of which are both the dynamic assignment of traffic and the optimisation of each individual element of the system (the network).

Indicators

The identification of saturation, as defined, implies the detection of abrupt changes in certain conditions which can be taken as indicators of the situation. These indicators are strongly correlated with the negative effects and can be listed in the same way :

For the off-the-road environment :

a) exhaust pollution level,

b) noise level,

c) amenity of the family dwellings in the area.

For pedestrians and cyclists :

a) accident rate,

b) number of impeded crossing opportunities per unit of time,

c) pedestrian accident indices.

For the in-vehicle trip-maker :

a) low speeds,
travel times corresponding to the above speeds,

b) excessively long travel delays,
intersection delays in excess of one cycle,
queue length at intersections,
number of side street blockings.

c) number of stops per link,

d) high densities,

e) high volume/capacity ratios.

II.2 SATURATION SITUATIONS AND PARAMETERS

The specific objectives to be pursued in a particular instance and the level of the approach used to reach them are to be weighed up by the professional who has to evaluate the effectiveness of the approach - and the efficiency of the means applied. This evaluation cannot be done on the basis of rigid patterns since each case is unique, depending on the amount of elements intervening in traffic (driver, vehicle, road, traffic), legislation, enforcement and a number of often random factors. Secondly, the approaches may vary due to the size of the problem-zone to be tackled, which may range from an isolated intersection to an area of five square kilometres. Furthermore, the level of operation may call for a broad scale of intervention, acting on the macrocauses, while in other instances the problem may be solved with microanalytical tools. Finally the main indicators and parameters may vary from case to case.

Consequently, this section considers the problem in units of increasing complexity, studying in each case the corresponding control objectives, parameters and measures of effectiveness.

Throughout the following assumptions will be made :

- the minimum study unit is the intersection ;
- the maximum study unit is the interurban access network of a populated area, or at least the area influenced by a populated area ; the rural highway or freeway network will therefore not be considered ;
- the study unit for traffic is the vehicle.

The possible saturated situations, together with the possible useful descriptive parameters, will be listed and commented on. In the next section the appropriate control measures for each will be described.

In order of increasing complexity, four cases may be highlighted.

1. Saturated Intersections. The individual intersections will be studied as a unit, disregarding the different geometric possibilities. Defining the intersection as a part of a road shared simultaneously by two or more cross-traffic flows, two cases may be studied :

   1.1. Internally Saturated Intersection. In this case the traffic volume of one of the intersection movements, for a significant period of time, is greater than the output allowed by the maximum phase the control device, or system, can apportion to it (assuming, too, that the entrance flow control to the intersection is less restrictive than the output control).

   1.2. Saturated Intersection Access. In this case the intersection is regulated by a non-optimised fixed-time system which produces a saturated access while the other accesses are not saturated. In other words, for a significant period of time, the demand at one or several access points exceeds the maximum possible output of the access control so that a queue always exists, even at the end of a green phase. This saturation state can appear at one, several, or all the intersection accesses.

   In this case a useful parameter to quantify the saturation state is the flow/capacity ratio on the accesses. The variation of queue length with time ($\delta Lq/\delta t$) also seems useful. The study of its evolution gives an advance indication of the congestion state which allows preventative action to be taken.

2. Saturated Arterials. An arterial is a route of a substantial length, giving service both to local and through traffic, possibly containing several intersections, merging points and diverging points.
Its basic unit is the link,i.e. a section between two intersections which will be considered first.

2.1. Saturation of a Link. A link reaches saturation when its quality of service suffers an abrupt drop as a consequence of a slight increase in demand. This condition can be easily detected by the appearance of a critical, or excessive queue length at an intersection approach. This critical queue length has been established around 65-85 m from the upstream intersection, and constitutes the threshold beyond which upstream drivers start to sense a congested link. Another useful parameter for this case is the flow/capacity ratio.

2.2. Saturation of the Route. Considering as an arterial a main street that crosses a given sector of an urban nucleus, two cases may be considered :

2.2.1. Saturation of the area surrounding the arterial : In this case the regulation of the arterial is such that it stays uncongested over all its length, though at the risk of congestion in the vicinity. The latter can be :

1. Longitudinal Saturation : The congestion appears at the ends of the arterial, outside its segment of interest (at one or both ends).
2. Transversal Congestion : The congestion is created at one or more of the at-grade crossing streets.

2.2.2. Saturation of the arterial : After having attempted to optimise every intersection and traffic movement, some link of the arterial is oversaturated. This may be :

1. Congestion resulting from an excess of internal demand.
2. Congestion resulting from a drop in capacity due to traffic contingencies (accidents, public works, pedestrians, (peak) demand).

3. Saturated Urban Network. Depending on the size of the urban area under consideration, the network may be considered as a set of crossing streets forming a grid. The following cases may be cited :

3.1. Overall Saturation of the Area or Urban Network. The whole urban centre is considered, allowing for total (accesses and internal network) or partial congestion, in which case the following may occur :

3.1.1. Saturated approaches : The situation of saturated approaches (streets giving access to the populated area) can take two forms :

a) Some approaches saturated :
   1. With possibility of diversion routes, or interchange with less saturated accesses.
   2. Without the possibility of diversion routes.
b) All accesses saturated.

3.1.2. Internal congestion of the network with free-flowing accesses :

a) Sub-area or partial urban network saturated, which includes saturated conditions at a localised geographical area of a reduced size.
b) The total network saturated.

4. Other Saturated Networks. This concept covers the system of roads partially or wholly outside an urban centre. It comprises two cases :

4.1. Saturated Accesses. Exterior roads giving access to towns or cities, which can be subdivided into the same cases as for the urban networks under paragraph 3.

4.2. Roads in the Vicinity of Towns or Cities. This may represent an interurban saturated flow. There may be three negative aspects :

1. Saturation produced by shock waves originating in the urban area.
2. Saturation produced by isolated capacity drops (accidents, ramps, turns, etc.).
3. Link saturation, produced by excessive inputs at upstream accesses.

Caution is needed in determining the specific sub-title to which a problem may be allocated. Avoidance of over-elaborate action measures is in line with the economy principle and, less far-reaching solutions should, therefore, be examined first.

II.3 POSSIBLE ACTION MEASURES

In accordance with the guidelines established by the Group, the different parameters and action measures for each case are listed below, with emphasis on the microanalytical approach. Measures dealing with increases of physical capacity or those acting on the demand outside the strict oversaturated area are given brief attention only. This Chapter is concerned solely with general guidelines. Chapters III, IV and V describe more detailed applications. Table 3 summarises the instances, parameters and possible action measures at the level presented in this chapter.

Measures can be classified as follows :

1. Saturated Intersection.

1.1. Internally Congested Intersection : The possible action measures to solve this instance of oversaturation are the following :

a) Realignment of the intersection to avoid mutual interference of different traffic flows at the approaches.
b) Reassignment of the regulation system's timing, (cycle length and split) and the intersection movement structure. This measure can reduce congestion up to a certain demand level.
c) Above this demand level, limitation of the intersection input at the accesses must be sought ; this limit will be the maximum possible output for the specific movement considered.

1.2. Intersection with Saturated Accesses : The usual action measures, which must be carefully assessed in each particular case, are the following :

a) Reassignment of the control system timing, in order to improve the service level of the specific access without saturating others of the same intersection.
b) Reducing cycle length if the spillback of the approach blocks side-streets.
c) When these measures are ineffective due to the demand volumes, evaluation of the saturation consequences for the whole set of accesses and allocation of the saturation among the different accesses so that all queues exceed their system boundaries at the same time ; in this evaluation the various hierarchical levels of the accesses and the type and composition of their traffic must be taken into account.

d) Reducing extra phases by other traffic options such as turn bays applying, shorter cycle lengths to create more end-of-cycle turning opportunities per hour, etc.

2. Saturated Arterials.

2.1. Saturation of a Link : The most usual parameters for this case have already been mentioned in Section II.2. The possible action measures are the following :

a) A capacity increase at the access exit by means of a reassignment of the intersection timing (cycle, split).
b) Changing offsets to favour progression, allowing for the discharge of residual queues.
c) Restraining the input flow, to a value below the queue capacity of the link, at the upstream merging point.
d) Increase of the number of lanes (reversible lanes, etc.).
e) Allowances for turn-on-right without extra phasing so as to reduce waiting queue.

2.2. Saturation of a Route : The same criteria as in the link case apply here, since an arterial is a composition of links. The following may, however, be added :

a) Coordination of the regulation of the various intersections, following progression criteria.
b) Queue distribution - breakdown - among the various arterial links to avoid cross-street blockage as far as possible.
c) When this is no longer possible, action measures will have to be at a second level, i.e. traffic assignment (re-routing), or in the long-term at the first level (land use, zoning, etc.).

3. Saturated Urban Networks. The applicable measures here include all the above plus the following :

a) Those on the macro-micro or second level : Traffic reorganisation, diversion routes, route allocated to specific traffic, etc.
b) Macro-control acting on land use, traffic generation or modal split, stimulating the traveller to use more efficient modes.
c) Finally, when none of these measures proves effective, the only appropriate measure is to aim at a fair distribution of saturation among the various routes and flows, taking here the delay as the unit of measurement.

4. Other Saturated Networks. The parameters in this case are the same as before, and primarily queue length.

When saturation occurs, the available strategies are those presented in II.3.2 and II.3.3. However, given this type of network with long links and low density of intersections, regulatory measures are quite useless above certain demand levels. Measures of the non-signal type may prove more useful, such as those mentioned in II.3.3.

Table 3

CONDITIONS, PARAMETERS AND ACTION MEASURES

<table>
<tr><th></th><th>Conditions</th><th>Parameters</th><th>Action measures</th></tr>
<tr><td rowspan="2">Saturated intersection</td><td>Internally saturated</td><td>Queue length
Flow/Cap, ratio</td><td>Realignment of operations
Reassignment of timings
Traffic limitation</td></tr>
<tr><td>Saturated intersection approach</td><td>Queue length
Flow/Cap, ratio</td><td>Reassignment of timings
Reduce cycle length
Reduce phases
Equity of adverse effects</td></tr>
<tr><td rowspan="2">Saturated arterial</td><td>Saturated link</td><td>Queue length
Flow/capacity ratio</td><td>Reassignment of timings
Phase arrangements to aid progression
Offset optimisation
Turn on right (or left) aids
Lane arrangements (one-way, reversible, etc.)</td></tr>
<tr><td>Saturated Route:
surroundings saturated: longitudinal saturation; transversal saturation
internally saturated arterial</td><td>Queue length
Flow/cap, ratio
Travel delay</td><td>Signal co-ordination
Queue distribution
Traffic assignment policies</td></tr>
<tr><td>Saturated urban networks</td><td>Area overall saturation:
Saturated approaches
internally saturated networks</td><td>Queue lengths
Flow/Cap, ratio
Travel delays</td><td>All above
Assignment policies
Modal split policies
Equitable distribution of adverse effects</td></tr>
<tr><td>Other saturated networks</td><td>Saturated accesses peripheral roads</td><td>All above</td><td>All above</td></tr>
</table>

III

STATE OF THE ART

III.1 INTRODUCTION

The purpose of this Chapter is to review the state-of-art of traffic signal control as it relates to congested or saturated conditions. Some general discussion of what "congestion" is follows. Such a state-of-the-art review should, of course, examine the existing practical solutions and recommendations. Perhaps the most important work in this area to date has been a study entitled "Traffic Control in Oversaturated Street Networks" undertaken in the United States for the National Cooperative Highway Research Program by the Polytechnic Institute of New York (NCHRP 194). This study defined the following terms :

1. Congested Operations - The entire range of operations experienced when traffic demand approaches and/or exceeds the capacity of the signal and queues are formed.
2. Saturated Operations - A subset of congested operations where the queues which form affect traffic performance only at the intersection at which they exist.
3. Oversaturated Operations - The other subset of congested operations where the queues which form fill entire blocks and interfere with the performance of adjacent upstream intersections.

For purposes of this report, the general term "congestion" will be used throughout to signify conditions of saturation.

The study mentioned also included a survey of traffic engineering professionals in the United States in an attempt to determine the root cause of congestion with the following results :

1. Lack of available alternate routes.
2. Land-use policies which generate congested traffic patterns.
3. Vehicle-pedestrian conflicts which interfere with traffic flow.

It is recommended that preventative actions which address these root causes be given top priority in treating the problem of congestion, such as :

- Preferential treatment of high-occupany vehicles.
- Staggered work-hours to spread vehicle demand over a longer period of time.
- Grade separation of pedestrian crossings.

Of more immediate interest to the subject of this report, the study recommends that traffic signalisation measures also be considered as a means of either preventing or alleviating the problem of congestion.

Congestion problems are usually local in nature at their onset, involving either a critical intersection or a specific link of limited capacity such as a bridge or tunnel. The problem may be temporary, such as occurs during a freeway incident, or may occur regularly, such as during peak-period congestion. If demand continues at or above capacity, congestion will spread and symptoms will become evident throughout the remainder of the arterial or network, involving completely loaded links and spillback through and beyond upstream intersections. Since the problem occurs at a critical intersection where demand exceeds capacity, the emphasis of this report will be on signal control strategies which can be applied at these critical locations to prevent congestion from spreading to upstream intersections and/or to manage the direction and rate of spread of congestion.

Signal control strategies, designed to deal with undersaturated flow conditions will differ greatly from those which are designed to work under congested conditions due to the different objectives to be satisfied.

Under moderate flow, the objective of the control is usually either the minimisation of vehicle delay or maximisation of bandwidth. Under saturated flow, the objective of the control shifts to the management of the growth of queues at specific critical locations where congestion is occurring. Failure to recognise this shift in the objective of the control can aggravate the congestion problem. By increasing throughput upstream of critical locations, even longer queues may be accumulated at these locations.

As noted in the NCHRP report, the signal timing parameters which are available to the traffic engineer are :

Signal split
Cycle length
Phasing
Signal offset

when utilising traffic signal control techniques.

One should, however, point out that the excess demand has to be accommodated somewhere. The NCHRP report states that the best use of all the links available should be made. Another not too different approach uses preselected links only (see Bordeaux). A third possibility is the gating or metering at the entrance to the network, a method that will often call for diversion schemes.

Having attempted to identify the root causes of congestion, the following sections of this Chapter contain descriptions and discussions of some of the more important techniques which are available to be applied to the problem of traffic congestion. Specifically the following items are included :

1. the control parameters and modes ;
2. computer-controlled systems ;
3. off-line optimization ;
4. some specific applications ;
5. theoretical strategies.

In the interest of completeness a number of other traffic engineering measures (non-signalisation) as identified in the NCHRP report are also included.

The final two sections address special problems of congestion in the areas of public transport and inter-city corridors.

III.2 TRAFFIC CONTROL PARAMETERS AND MODES

In attempting to solve problems of traffic congestion, the traffic engineer may employ several methods of control, and in applying these methods may manipulate several parameters.

1. Controllers utilise several methods of operation such as fixed-time, traffic-responsive, or centralised computer-based control.
2. Signal timing plans may be developed on or off-line and may be based on optimisation procedures.
3. The basic control parameters, cycle length, signal split, signal offset, and sequence of phases, may be manipulated in applying the control.

III.2.1 Type of Control

Under moderate flows, controllers which operate in the traffic-responsive mode can be shown to be much more effective than fixed-time controllers. At or near saturation, however, the performance of the two modes may be similar, in fact in many cases actuated controllers will operate as a fixed time-controller under congested conditions (see Figure 4).

Figure 4 **TYPICAL DELAY-FLOW CURVES FOR VEHICLE ACTUATED AND FIXED TIME SIGNAL CONTROLS**

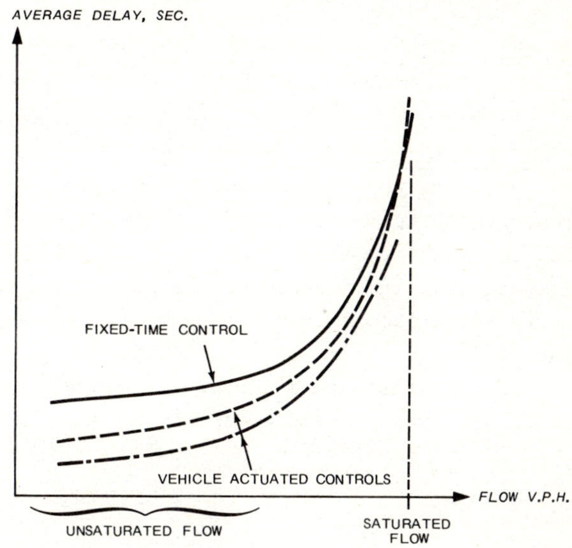

The advantage offered by actuated controllers is quick response to random variations in normal traffic flow patterns (see Appendix B). Many congestion problems, however, are peak-period problems which are repeated on a daily basis. Under these conditions, fixed-time controllers which operate with adequate time-of-day or computer-controlled monitoring and updating capabilities are efficient. Depending upon the situation, actuated controllers may offer an advantage in situations where a random occurrence such as a lane blockage causes congestion at an isolated location. In such cases, an actuated controller could respond to the situation and adapt the signal timings accordingly.

III.2.2 Calculations of Cycle Length

The findings of the United States study indicate that contrary to the common belief that longer cycle lengths result in increased capacity and therefore should be employed in dealing with congestion, the small gain in capacity is offset by the disadvantages resulting from increased delay and longer queues.

As an example, using Webster's methods to determine cycle length, it can be shown that to achieve even a modest increase of 3 % in vehicle flow, it would be necessary to increase the cycle length by 70 % which is usually impossible to achieve in practice.

Cycle length should be constrained by the amount of storage available on the upstream link. The length of queues should not be larger than the available link storage. This may be expressed :

$$f\left(\frac{C}{3600}\right)d \leq L$$

where,

f = flow in vph,

C = cycle length in secs.,

d = average vehicle storage length (metres per vehicle), and

L = available link storage (metres).

Rearranging the expression results in the following for maximum permissible cycle length :

$$C = \left(\frac{3600}{f}\right)\frac{L}{d}$$

Cycle length may also be constrained by the amount of storage available on the downstream link. Appropriate values for f and L in the above equation will ensure that the flow being discharged from the critical intersection will not cause congestion to occur at the downstream intersection.

Another reason for short cycles is the fact that when there are no protected phases (United States), or when there are secondary conflicts (Europe), the capacity depends on the number and not on the length of the cycles.

III.2.3 Calculation of Green Signal Time (Split)

The NCHRP report also states that the standard procedures for determining signal split (i.e. signal split is proportional to the ratio of critical lane flows or critical lane flow to capacity ratio) are not satisfactory in oversaturated conditions. These methods do not consider the effects of spillback when the intersection approach volume exceeds capacity. The presence of spillback limits the critical lane flow which would indicate the need for less green time on that approach which is the opposite conclusion that should be reached. A procedure is recommended to ensure that the rate of growth of congestion in all directions is equalised, i.e. spillback on all links should occur at the same time.

$$g_A = \frac{(f_A/S) + (Q_A/Q_B)(K-f_B/S)}{(1 + Q_A/Q_B)}$$

where,

g_A = proportion of cycle for green on Phase A

f_A = per-lane demand on Phase A in vph

S = per-lane saturation discharge in vph

$K = 1 - (L/C) = g_A + g_B$

$Q_A = (L_A - f_A C)$

L = loss time per cycle in secs.

C = cycle length in secs.

L_A = available storage per lane in number of vehicles

If congestion is such that spillback beyond the upstream intersection is inevitable, the above equation can be used to limit the total extent of spread by substituting the absolute distance beyond which spillback is not permissible as the value of L_A.

III.2.4 Calculations of Offset

During periods of heavy flow, forward progression is obviously undesirable since such action contributes to the formation of queues creating spillback on congested links. For congested conditions offsets should be adjusted to allow queues which have formed to discharge before additional vehicles arrive. The NCHRP report includes an equation which can be used to determine this offset.

$$t = \frac{L}{V} - \frac{Q}{R}$$

where,

t = desired offset in secs.
L = signal spacing in metres
V = design speed in meters per second
Q = estimated queue length in vehicles per lane, and
R = discharge rate in vehicles per second.

This equation may be applied to all conditions, since as congestion increases the offset is reduced in value and green phases start earlier. When zero or a negative value is reached, simultaneous or reverse progression is attained.

When spillback is inevitable, offsets can be calculated which ensure equitable treatment of competing flows upstream of the critical intersection. Thus, the basic concept of what the offset accomplishes is changed. The following equation is contained in the NCHRP report :

$$t_{OFF} = g_1 C - \frac{L}{V_{ACC}}$$

where,

t_{OFF} = desired offset in secs.
g_1 = proportion of green time at the intersection upstream of the critical intersection
C = cycle length in secs.
L = signal spacing in metres, and
V_{ACC} = speed of the acceleration wave in metres per second

Figure 5 **USE OF OFFSET TO ENSURE EQUITABLE TREATMENT OF COMPETING FLOWS AT THE UPSTREAM INTERSECTION**

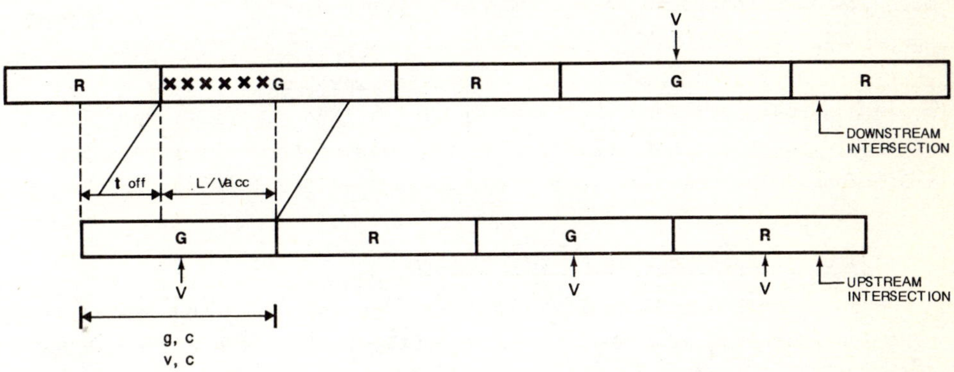

Through the use of this offset between the critical intersection and its upstream intersection, the oversaturated through-movement is permitted to have green at the upstream intersection until vehicles blocking it just begin to move. Green is then switched to the cross-street traffic and this movement is accommodated until it has had equitable access. Care should be taken to ensure that the cross movement has had fair access, but input from the cross movement should not fill all of the available storage space since this would block the through movement and congestion would propagate upstream.

This concept is recommended for use only as a last resort when other measures have failed and congestion is unavoidable.

III.2.5 Signal Phasing

As a general rule, the NCHRP report recommends that multiple phases be avoided since these generally require an increase in cycle length and result in more lost time per cycle. Instead, it recommends that the following options be considered :

1. Installation of turn bays, without protected phases, be used to alleviate turning movement problems.
2. Shorter cycle lengths to provide more end-of-cycle turning opportunities per hour.
3. Parking restrictions near the intersection on single-lane approaches to provide by-pass opportunities for through traffic that would be otherwise impeded.
4. Leading and/or lagging greens to aid turning vehicles, and
5. Turn prohibitions if alternate routes are available.

Strictly speaking these are not traffic light control measures, but they are very important because they often remedy the whole saturation problem by solving the critical intersection problem.

III.3 COMPUTER-CONTROLLED CONTROL SYSTEMS - (ON-LINE)

A number of computer-controlled traffic signal control systems have been developed. The operational features of the most significant of these systems are described in Appendix C. Computer-based signal operations offer a number of features which are not available with conventional hardware :

1. Perhaps the most significant advantage is the provision of many more timing plans for specific time-of-day operations.
2. A computer-based system can operate in a traffic-responsive mode, accepting measured traffic demand through some surveillance mechanism, and selecting a timing plan from a library of plans stored in the computer. This selection process is achieved by using a pattern-matching algorithm or threshold logic. Updating most commonly occurs in fifteen minute intervals.

Computer-based control is thus better suited to provide timing plans to match the existing traffic conditions. Prevention of congestion may not be possible however, due to the time involved between plan updates, and the limitations of traffic prediction algorithms.

The following sections note some of the features of the various systems described in Appendix C, emphasising the aspects of each which specifically address the problem of congestion.

III.3.1 UTCS - Urban Traffic Control System

The UTCS as described in Appendix C was designed in the United States to develop advanced operational control programs. Three generations of software were investigated as were specific procedures to deal with congested conditions as described below.

First Generation software involves the use of prestored timing plans which are selected on the basis of prevailing conditions (time of day, traffic flow, or operator's choice). Critical intersections are identified on the basis of frequency of saturation at these and green times are adjusted once per cycle as a function of local demand. In this way the local control supplements area control to alleviate local congestion.

Second Generation software involves a real-time on-line system which computes and implements timing plans based on surveillance data and predicted changes in demand.

A Critical Intersection Control (C.I.C.) routine changes splits and may change offsets once per cycle as a function of the local demand (at critical locations). The changes made depend upon the degree of congestion detected as defined by pre-set thresholds. Because C.I.C. has been found to perform poorly when large random volume fluctuations exist, no changes are made for undersaturated or oversaturated unstable conditions.

Third Generation software has been conceived as a fully responsive on-line control system. The design will permit cycle length to vary from cycle to cycle with appropriate offsets and splits to minimise delays and stops.

Both congested Intersection Control and Queue Management Control (C.I.C./Q.M.C.) are included in third generation design to deal with congested intersections or links. The objective of this strategy is to manage the congestion within a restricted area which is under C.I.C./Q.M.C. control.

C.I.C. is applied at those intersections when they become saturated. Q.M.C. is concurrently applied at the two upstream intersections of the dominant approaches. At the critical intersection, signal split is apportioned to the competing approaches based on the extent of congestion on all links along the competing congested paths. Q.M.C. is deployed at nodes upstream of the congested intersection in order to manage the queue length on links which lie immediately downstream on the path to the congested intersection. Q.M.C. attempts to provide signal co-ordination which will maintain a minimum queue on the downstream link at the onset of the green indication at the downstream node.

It should be noted that C.I.C./Q.M.C. control procedures have not been adequately tested or evaluated.

III.3.2 PBIL - Responsive Control in Aachen, Germany

The PBIL control system described in Appendix C was developed as a directly responsive concept and was tested on-line over a limited number of functions in Aachen, Germany. The system directly addresses congestion problems and is designed to manage a critical intersection and the adjacent intersections with which it interacts. The steps in the procedure are :

1. Identification of Critical Locations.
2. Separation of these locations from the balance of the system.
3. The operation of these areas under PBIL.

Basically, the system tries to reduce total journey delays. However, when an arm of the intersection gets nearly saturated, the system acts as if delays thereon are much higher than in reality. With this "false" information the system will automatically try to prevent the congestion problem.

III.3.3 SCOOT - TRRL - Application in Glasgow, United Kingdom

SCOOT is a traffic responsive system developed jointly by TRRL and the British traffic system companies and tested in Glasgow (50 junctions) and also in Coventry (27 junctions). SCOOT uses an on-line model of traffic on all links in a network and optimises splits, offset and cycle times in a hierarchy of optimisers. The philosophy is make frequent small changes in order to avoid plan change type transients and to react quickly to random cyclic fluctuations, as well as trends in flow level.

Special procedures are designed to deal with congested conditions. Queues can be measured on each link and when congestion is imminent, the split and offset optimisers will take account of this in their performance indexes and favour congested links. Green time may be limited upstream in the event of spillback. In these ways SCOOT provides a "preventative" type of control. SCOOT is not, however, designed to control a completely oversaturated network.

III.3.4 SCAT - Sydney Co-ordinated Area Traffic System, Australia

SCAT is a traffic-responsive system designed to provide both local (tactical) control and area (strategic) control (see Appendix C). In light traffic conditions, SCAT attempts to minimise stops, in medium traffic conditions, to minimise overall delay, while in congested conditions splits and offsets are designed to maximise throughput on major routes.

A segment of the program recognises the onset of congestion through detector inputs, and alters the splits to minimise the adverse effects of congestion (queues are stored on links where their presence is least likely to cause system failure). Concurrently offsets are altered to produce maximum throughput and full blocks.

III.3.5 GERTRUDE - Bordeaux, France

The computer-based control program know as GERTRUDE was designed specifically to deal with congested conditions and has been put in operation in Bordeaux, France (see Appendix C). The system operates with a fixed emergency routine to detect congestion and to initiate preventative plans (as soon as congestion is expected) over specific areas, each comprised of critical points and the neighbouring intersections.

In order to prevent spillback the program rearranges signal splits to maximise throughput at the critical intersection. This action is taken to avoid blocking upstream intersections at the expense of particular links (storage links). These links are selected to serve as buffers which receive all excess demand in the main approach roads.

III.3.6 R.T.O.P. - Realtime Optimisation Program - Toronto, Canada

This program was designed as a second generation system to supercede the computer-based system which has operated in Metropolitan Toronto, Canada, for many years (see Appendix C). The system design comprises three levels of control : local control (micro-processor type intersection control), celluler control (tactical) and system-wide control (strategic).

No major component of the system has been provided at this time to address the problems of congestion. However, individual intersections with a history of congestion can be isolated and operated as traffic-responsive units when conditions dictate such action.

III.4 OFF-LINE OPTIMISATION TECHNIQUES

In order to determine values for signal-timing parameters to be implemented in the field, many traffic engineers now make use of optimisation programs to generate timing plans off-line.

The program may operate on the basis of either :

a) Maximising the Bandwidth or

b) Optimising a combination of traffic Measures of Effectiveness (M.O.E.) such as stops and delays.

Some of the more commonly used programs of this type are outlined in Appendix D. However, the use of such techniques to provide timing arrangements to be used in congested areas are suspect as discussed in the following sections.

III.4.1 Bandwidth Programs

Bandwidth programs (e.g. SIGSET, PASSER II, MAXBAND) operate on the principle of maximising the width of the band of vehicles which is able to successfully encounter green signals when travelling along a street. These programs are applicable to arterial streets, in particular on arterials having insignificant cross-street flow. Such programs attempt to maximise bandwidth on the main street alone and only the split calculation is involved with respect to the cross street. Bandwidth programs are not applicable to congested conditions because they contain no provisions for queue management nor do they consider contributing flow from the cross streets. Even in cases where a provision to initiate green early at an intersection on an arterial to provide for queue clearance, queue buildup or spillback is not considered.

III.4.2 Optimisation Models and Measures of Effectiveness

Optimisation Programs operate on the principle that the value of one or more network-wide traffic MOE (measures of effectiveness) should be optimised (e.g. a linear combination of stops and delay is minimised). The programs are applicable on both arterial streets and in urban networks. Two of the more widely known of these programs are the Signal Optimisation (SIGOP II) program developed in the United States and the TRANSYT program developed in Great Britain. These programs are described in Appendix D which also discusses the shortcomings of these programs when applied to congested conditions.

TRANSYT

This program, developed by TRRL, United Kingdom, attempts to optimise a value known as the Performance Index. This index is the weighted amount of total delay and stops in the network. Traffic flow is represented by a simple but effective model which essentially uses an upstream to downstream process to describe the traffic characteristics. Traffic flow is modelled over a one-cycle duration as a unit.

Although the TRANSYT Program has proved to be very effective and is rapidly growing in popularity in many countries, it nonetheless has some shortcomings when applied to congested conditions for which it was not intended.

1. Since the traffic is modelled upstream to downstream over one cycle, and the resulting discharge from an upstream signal becomes the input for a downstream signal, vehicles which could not clear the downstream intersection (congestion) and are therefore stored in the upstream link are not accounted for since succeeding cycles are not modelled.
2. The additional delay associated with these excess vehicles is not fully accounted for in the random delay calculations.
3. On the downstream node a miscalculation of the random delay is made as all vehicles are expected to arrive.
4. The model does not recognise the physical extent of the queue that is thus forming from the ever increasing excess vehicles (spillback problems are therefore not anticipated).

For these reasons this program in its current form cannot accurately model congested conditions. (Note : Officials in the Netherlands have made some modifications to the TRANSYT model to enable it to deal with congested conditions. Similar improvements are now incorporated in the new TRANSYT 8 version of TRRL).

SIGOP II

The SIGOP II program, developed for the Federal Highway Administration in the United States, is described in Appendix D. It is comprised of a model of traffic behaviour and an optimisation program, and is very similar to the TRANSYT Program with the following differences :

1. The SIGOP II performance index is a weighted combination of stops, delays, and queue length.
2. The traffic model in SIGOP II is somewhat cruder than TRANSYT's, particularly in how platoon behaviour is replicated.
3. The optimisation scheme in SIGOP II attempts to minimise a combination of MOE's expressed as a quadratic equation.

The shortcomings of the SIGOP II program in relation to congested conditions are basically the same as for the TRANSYT program. SIGOP II also models flow over one cycle ; therefore, delay is not accumulated and spillback can not be considered.

SIGOP II, however, contains the following provisions to account for the effects of congestion :

1. In cases where an intersection is saturated, SIGOP II will attempt to provide more green time to the saturated approach.
2. The stops and delay calculations (which serve for the performance index) are modified to better account for the effects of congested flow by penalizing excessive queues.

III.5 SPECIFIC APPLICATIONS

A review of the literature indicates that few applications of strategies specifically designed for control problems in congested conditions have been reported to date. Many solutions to individual problems have no doubt been successfully used but not documented. Some specific examples are described below.

1. A technique is used in France and Japan to prevent blockage of an upstream intersection due to an excessive increase in queue length from a downstream intersection. The signal facing the congested path at the upstream intersection is switched to red when occupancy of a protection loop, located near the point of maximum queue from the downstream intersection, exceeds a certain value.
2. A special all-red period is sometimes employed in France and in Ireland to protect against turning vehicles blocking an intersection, detected by means of an inductive loop or radar detector. On British controllers an all-red phase is called and extended by a mid-junction detector and has minimum and maximum red settings.
3. The city of Rotterdam (the Netherlands) combined inductive loop detection on approaches with speed measurement at the same spot using overhead ultrasonic speed detectors to provide quick response when the maximum allowable queues are quite short. The controller responds by increasing the appropriate green time, or by blocking traffic coming into the link.
4. In a few instances in the Netherlands cycles are utilised which display the green indication twice. This brings more opportunity for queues to discharge before reaching critical lengths.
5. In the city of Hamilton, Ontario, Canada, the extensive use of one-way streets together with fixed time signals create almost perfect vehicular progression which is highly successful during periods of undersaturated flow. During the periods of peak flow, however, the turning movements at upstream intersections tend to fill the storage space on certain arterial links approaching critical intersections. In such cases, special splits and offsets have been introduced which no longer favour an uninterrupted through movement, but rather attempt to maintain clear storage space on the critical links and thus minimise the effects of spillbacks.
6. In the city of Winnipeg, Manitoba, Canada, two 2x4-lane arterials form a major intersection in the central business district. In order to serve the major traffic flows and provide adequate pedestrian crossing times, a multi-phase operation has been provided and hence available green time for each approach is limited. Most of the city transit buses traverse these routes and encounter extreme difficulty in changing lanes prior to completing turning movements due to the presence

of extensive queues on the links approaching the critical intersection and cause queues to form at other upstream intersections.

7. In Yokohamo, Japan, cycle-by-cycle on-line feedback control was employed to control the traffic signals in a congested network. Average pulse length detected at an appropriate location was found to be the best indicator of the degree of saturation of intersection approaches. Micro-computers were used to process detector pulses as well as control split. The control strategy attempted to keep the degrees of saturation of all of the signal phases as similar to each other as practical.

III.6 OTHER TRAFFIC ENGINEERING MEASURES (NON-SIGNALISATION)

In many cases, traffic signalization measures alone are not sufficient to solve the problems of congestion. Taking certain steps in terms of traffic management will often enable more effective use of the proposed traffic signal strategies - the signals are able to do the job intended.

Although not strictly within the terms of reference of this report, nonetheless the traffic engineer, in attempting to design remedial measures for congestion, will of necessity be required to consider one or more of such related measures. A list of some of these associated strategies and/or policies are listed below (see NCHRP Report No. 194) :

1. Parking prohibition provides additional lanes, to increase the capacity of the facility or to provide a special turn lane or bay.
2. A stricter enforcement policy for double parkers and drivers blocking intersections.
3. Turn prohibitions where alternate routes are available and congestion is caused by delays due to turning vehicles.
4. Provision of turn bays to separate turning traffic from through traffic, thus increasing the effective green time for through vehicles.
5. Provision of two-way turn lanes (i.e. a lane that can be used by vehicles in either flow direction waiting to execute their turns) when frequent mid-block turns are causing a congestion problem.
6. Provision of dual-turn lanes when turning volumes are extremely heavy.
7. Provision of one-way street systems, where feasible.
8. Reversible lanes, in which one or more lanes are assigned to one direction or the other depending upon the time of day.
9. Separation of pedestrian and vehicle traffic through the use of pedestrian vehicle separation (bridge or tunnel).
10. Provision of bus stops which are separated from traffic flow or relocation of bus stops away from critical approaches.

III.7 THEORETICAL STRATEGIES

This Section outlines some of the theoretical strategies which have been proposed to deal with congestion. Details of the mathematical developments are to be found in Appendix E.

III.7.1 General Discussion

As has been stated elsewhere in this report, solutions to problems of traffic congestion are related to the critical intersection. Our first concern therefore must be to consider the individual intersection as the basic unit and more particularly the individual approaches to the intersection in the form of upstream and downstream links. After measures based on an intersection analysis have been considered, metering techniques as a second type of measure can be applied.

A truly isolated intersection is one where the queues obtained do not interfere with an upstream intersection. The control objective for such an intersection would be to minimize average delay for all arrivals within some time interval. In most practical cases, however, queue lengths also have to be considered. This can theoretically be taken care of by introducing constraints into the control policy. Other strategies consider minimisation of queue length to be the objective or attempt to maintain a constant ratio between queue lengths on the critical approaches of each signal phase.

The task then is to devise timing plans which will :

1. apportion green time between the competing traffic demands at the critical intersection in such a manner as to minimise overall vehicular delay.
2. manage the queues on the associated links in such a manner as to avoid spillback into adjacent intersections and/or to equalise the queues on each link.

Various researchers have attempted to find theoretical solutions to these problems utilising both off-line and on-line techniques as discussed in the following sections.

III.7.2 Off-Line Models

The selection of timing plans for a congested intersection by means of an off-line model can be envisaged as a generalisation of selecting time settings for a fixed time signal with constant demand. If congestion is present over an extended time frame, delay and queue lengths will be more sensitive to system-wide time variation in demand than is the case during normal operation. It may be useful to calculate time settings as a function of expected demand (time-dependent).

The following model developments can be described :

1. The determination of time-dependent settings for a fixed time signal with behaviour demand variations was first treated by Gazis and Potts. Their objective was to minimise the total aggregate delay at an intersection. Using queue lengths as state variables and service rates as control variables, the problem is solved with the use of Pontryagin's control theory. In general, the optimal policy consists of a two-stage operation ("bang-bang")*. During the first stage, the approach with the highest saturation flow is given maximum green time and the other approach minimum. At some point when the queue in the first approach is exhausted, the setting are reversed. The choice of switch-over point that minimises aggregate delay, however, does not in general minimise the period during which there will be congestion on at least one approach.
2. By dividing the time interval considered into periods of one-cycle lengths, the problem is made amenable to linear programming techniques. Using the Simplex method, d'Ans and Gazis found the time-dependent green time values using the same objective as before. Church and Revelle introduced maximum waiting time for any one vehicle and maximum queue length as constraints in the problem formulation.

   Michalopoulos and Stephanopolous performed similar work and also studied the effect of the queue-length constraint. It was shown that the constraint had no

---

* Expression for a control mode that switches from one extreme solution to another.

effect until the limiting values were reached. When this occurs, the optimal control will be to balance input and output rates, maintaining a constant queue length until some switch point. Church and Revelle also considered the use of maximum queue length as the objective instead of aggregate delay. The solution tends to balance queue lengths on the most saturated approaches of each signal phase, with frequent switches between phases.

3. Gazis expanded his model to a simple system of two nearby intersections. His work was further expanded by Burhardt, who included the effect of travel time between intersections, and by Michalopoulos, who also considered a maximum queue-length constraint. The solutions obtained are still "bang-bang". At any point in time, the green time at any approach will be either minimum or maximum.

An essential feature of the models mentioned is the deterministic view of the traffic process which makes the control problem analogous to that of reservoir management. When the time horizon considered includes periods where capacity exceeds demand on at least one approach, stochastic effects should be included. This unfortunately makes the problem non-linear and much more complex. Yagar has utilised a continuous delay/demand relation, including a stochastic term when capacity exceeds demand to optimise splits. An interesting future development could be the treatment of random arrival effects as linear random constraints, resulting in a linear programming formulation with uncertainty.

III.7.3 On-Line Models

On-line strategies are considered to be theoretically less effective under congested conditions, in that future flow-rate information is not utilised. As the extent of congestion decreases however and flow rates become less predictable, the effect of random variations will be more important and on-line strategies more effective. On-line control may also be required at locations with critical queue-length constraints when even short term excesses cannot be tolerated.

On-line strategies which attempt to minimise delay for a congested intersection have been proposed by many researchers. Their approaches are similar using detector outputs (analysed frequently) to select a control policy predicted to minimise delay during the next signal cycle. For example, Ross et al assume that the controlled intersection is the critical one in a network and the settings at the nearest downstream intersections are considered. The effect on total delay of different discrete extensions of the green period for an approach is estimated by a deterministic model. The number of expected arrivals is obtained from demand detectors far upstream. A series of simulation studies with demands up to 20 % above capacity for short periods resulted in considerable delay reductions when the above strategy was compared with fixed time control.

Other on-line models for saturated conditions aim primarily at reducing maximum queue length to prevent blocking of nearby intersections.

1. Gordon assumes one approach in each of the two signal phases of an isolated intersection to be critical and suggests a control that aims to maintain a constant ratio between the respective storage spaces. The input is the change in queue length between successive cycles and the output is the change in split between the cycles, assuming a fixed cycle length.
2. Longley attempts to maintain balance among the queues on all four approaches to an intersection in his control strategy. At the end of each cycle, split is determined for the next cycle based on the results of the queue balancing formulation for all preceding cycles.

III.7.4 Network Considerations

Some theoretical work has been undertaken with regard to overall signal networks :

1. d'Ans and Gazis proposed a policy which considers not only vehicle demand but also vehicle paths through the networks. The strategy is aimed at minimising aggregate delay in the network. Each link has associated with it a certain travel time, capacity, and storage space. Route distribution is assumed to be known as a function of time, resulting in known demand rates for each movement at each node. Discharge rates of the various O-D pairs on an approach are assumed to be proportional to the respective queue sizes. The resulting formulation yields a linear operational research problem only in the case of a single O-D pair. A method for practical application to the general case has not been developed.
2. Tödli utilises route-distribution information to control the amount of turning traffic at each node during a given time interval. The turning ratios would be dependent on the signal settings at the upstream node. An iterative technique is suggested whereby splits are obtained through linear programming for fixed turning ratios and are then adjusted based on current information. Tödli illustrates the technique for an academic example, consisting of a circular one-way road with approaches.
3. Tödli further suggests that this strategy be modified by on-line control, based on observed queue lengths, in order to correct for random demand variations. Proportional control is suggested where the change in green is proportional to the change in queue length.

III.8 PUBLIC TRANSPORT IN CONGESTED CONDITIONS

Of particular concern in attempting to solve problems of traffic congestion, are those encountered by Public Transport vehicles. Some of the strategies currently being used and associated working experiences are presented in this Section.

III.8.1 The Problem

When intermingled with general traffic, public transport vehicles are particularly vulnerable to congestion. A bus which is delayed by congestion will find more passengers waiting at the stops and will continue to lose time. A following bus will tend to find fewer passengers at the stops and may overtake the first bus resulting in the familiar phenomenon of bunching. In this way, service on the whole line is disrupted.

It is extremely important to avoid these disruptions both from the point of view of the users (irregular frequency of service and uncertainties as to travel times) and of the operators (inefficient use of the vehicle fleet, service delays). Measures to tackle these problems currently include the introduction of bus lanes, signal control measures, and the centralised management of public transport lines.

III.8.2 Bus Lanes

Buses may be routed along reserved lanes over parts of their journey where traffic problems are particularly acute. The OECD has published a report on bus lanes where more details may be found.

It is worth stressing that all the studies on the effectiveness of bus lanes have shown that they result in an improvement in the speed and regularity of service in periods of congestion. The degree of improvement has varied markedly from site to site.

Some limits to the effectiveness of bus lanes are :

a) Problems related to deliveries, and more generally to the failure to respect the bus lanes.
b) The problem of turns at intersections (i.e. left or right turners using or cutting across the reserved lane).
c) The volume of bus traffic in the lane. Low volume encourage lane violations and the operation is probably not beneficial in terms of overall passenger delay. On the other hand, beyond a certain critical flow the benefits of the lane tend to be negligible.
d) Delay at the stops along the lane when it is being used by several lines sharing the same stops.

It may not always be physically possible to install bus lanes along congested roads, particularly when pavement width is inadequate.

Good results may be obtained by "gating" methods such as those introduced in Bordeaux and Southampton. Bus lanes may be established at the entry points to the controlled zone in order to avoid bus delays. In Bordeaux, a special signal for buses at the exit points of the storage zone allows them to move to the head of the traffic queue. The Bitterne Road bus priority scheme in Southampton incorporates a number of regulatory measures : signal synchronisation, one-way streets, streets reserved for buses, and restrictions on turning. As a result of this scheme, the circulation of bus traffic has improved at the expense of additional delay to private vehicles.

III.8.3 Signal Control Measures

The time spent waiting at traffic signals represents a considerable proportion of total travel time, (as much as 40 percent according to a recent British study). Measures for making specific provisions for buses have been investigated extensively and several possibilities are discussed below.

a) Design of Fixed Signal Timing to Favour Buses

Buses can rarely take advantage of signal co-ordination along an arterial because they must halt at the stops between the signals. The TRANSYT 5 program was developed at the TRRL to overcome this problem. The number of stops and delay encountered by buses may be specifically accounted for in the performance index used to optimise the signal timings in the program. Bus speeds were increased by 8 per cent with no significant change to the speed of other vehicles.

b) Bus Priority at Intersections

One way of reducing bus (or light rail vehicle) delay at signalised intersections is to design a form of preemption to extend the green phase if the signal is green and if the bus has insufficient time to pass, or to truncate the red phase if the bus arrives when the signal is red, or to do nothing if the signal is green and the bus has time to pass. It is also possible to initiate a specific phase earlier in the cycle in order to grant priority to buses and thereby improve bus flow.

Results have shown that in the case of isolated intersections or where there is no route conflict, bus priority can reduce travel time, with adverse impact on the general traffic flow. In the case of conflicting bus routes, however, gains have been appreciably less and the system of priority at signals can even degrade bus flow when the frequency of arrivals is too high. Little research has been undertaken in congested conditions. Bus priority studies in Washington have concluded that it is preferable to disconnect the priority system in periods of congestion. Simulation studies undertaken by the French Transport Research Institute have shown that priority is always favourable to buses, even in peak periods, except in the case of conflicting bus routes. In particularly heavy traffic, however, it appeared that fixed signal timings for buses (TRANSYT 5) gave the best results for users as a whole (the criterion being travel time per user). Capacity studies in The Netherlands show that providing for bus priority requires excess capacity. For example, to provide a three-minute bus' interval, one needs about 10 percent excess capacity.

c) Other Signal Control Measures

In Sweden, a method for signal control at isolated intersections derived from Miller's algorithm but designed specifically for buses has been tested both in the field and by simulation. The results show that this method, known as TOL, is preferable to a fixed cycle or to regulation by critical interval in all traffic conditions, especially during peak periods. In this case, the bus got a weighting factor in the Miller delay optimisation model which equated the bus to 30 cars.

III.8.4 Centralised Management of Public Transport

Many cities are equipped with centralised management systems for public transport (e.g. Besançon, Chicago, Dublin, Hamburg, Hannover, Zürich, Toronto). All of these systems operate on a similar basis. Buses may be contacted periodically by radio from a control station and, in turn, will relay their position and other relevant data. The position of vehicles may also be located by means of detectors along the route or by counting the number of wheel revolutions from the terminus or a combination of the two methods.

a) Methodology

These systems are designed to immediately alert the operators to any disruption and enable them to take quick and effective action before the situation becomes serious (e.g.) by sending out a repair crew, bringing a new vehicle into service, police assistance). Most often these systems simply provide the operators with a picture of the state of the network. In certain cases (Besançon, Hannover, Zürich), however, instructions are relayed automatically to drivers (position vis-à-vis schedule, orders to leave a terminus or regulation station etc.). In Besançon, the system can also propose a set of responses for specific cases.

b) System Use With Bus Priority Scheme

Systems for centralised bus management may be used in conjunction with a central traffic signal control unit to manage priorities at traffic signals over the whole network rather than at each intersection, taking into account the state of the bus network (schedule delays, headways) and the general traffic situation. In this way it is possible to give priority to buses only when they are late, to use the signals to space buses more evenly or to eliminate priority measures when the general traffic flow is excessively disrupted. Such systems are in operation or being set up in Hannover, Lille, and Strasbourg. No data on their effectiveness is yet available.

III.9 INTER-CITY TRAFFIC CORRIDORS

Problems of congestion in traffic corridors connecting urban centres have received the attention of traffic engineers in recent years. This Section considers the various methods which have been developed for this purpose (see Appendix F for details).

III.9.1 Automated Corridor Control Systems

Implementation of an automated corridor control system benefits both the users and operators of the system. From the user's point of view, reductions are achieved in accident rates, travel times, and vehicle operating expenses. From the operator's point of view, reductions in road construction investments and maintenance costs are achieved.

The operation of these systems can be generally described in a series of steps as follows :

1. Traffic flow in the network is detected by loops placed in critical locations and traffic data are transmitted to a central control computer.
2. The traffic data gathered are processed and analysed.
3. On the basis of the traffic analysis, a short-term prediction, perhaps for the next 5 minutes, is made which gives detailed information on the imminent density at the critical locations and the relationships between density, volume, and speed.
4. Using the predicted traffic data, a performance index (PI) is calculated for each section. The PI may include estimates of travel time, fuel consumption, and accident probabilities.
5. Estimates of the PI are determined for alternative traffic guidance or access control strategies. The strategy which produces the optimum PI will be implemented. Route selection is controlled through either advisory information to the motorist or through access control measures.

III.9.2 Methods of Control

In the following, the state-of-the-art of traffic guidance and access control systems is reviewed (see Appendix F for details regarding application).

The various techniques may be summarised in terms of the technology employed to transfer information to the road user such as :

Broadcasting
Variable Message Signing (VMS)
Access Control
Integrated System
Individual Vehicle Guidance

In general these terms may be briefly described as follows :

1. Broadcasting. Information is dispensed to road users either before or during their journey as to existing or predicted conditions in the network. Drivers are thus assisted in making decisions concerning their route as indicated in Appendix F at F3. This method is widely used.
2. Variable Message Signs (VMS). Using changeable message signs, strategically located either in the network or on the approaches thereto, the traffic engineer can relay important information to road users. These messages may either warn of specific congestion incidents and/or provide information on alternate routes. As in the case of broadcast data, the individual trip-makers are assisted in choosing appropriate routes.
3. Access Control. This strategy involves limiting the amount of traffic which is permitted to enter the network. Ramp metering devices are employed at critical entry points to control the volume entering during a particular time interval. (Typical examples can be found in France, Germany, United States and Canada).

4. Integrated System. A system, known as the INTEGRATED MOTORIST INFORMATION SYSTEM (IMIS), is currently being developed in the United States. The control strategy is designed to provide cost-effective traffic management and congestion control in heavily-travelled urban corridors (see Appendix F).
5. Individual Vehicle Guidance. This technique involves providing individual guidance to motorists through on-board electronic aids. As indicated in Appendix F development work is underway in Germany and Japan.

III.9.3 Off-Line Models

A number of freeway off-line models have been developed in the United States and France, as outlined in Appendix F. These models have been designed for specific functions :

1. FREQ6PL is used for the evaluation of reserved lanes on freeways - United States.
2. FREQ6PE is used for the evaluation of priority entry, v.s. normal entry control - United States.
3. PERCEVAL is used to calculate metering rates for approach roads on the basis of an O-D matrix - France.

IV

CRITICAL ANALYSIS

Congestion creates traffic conditions which are more difficult to alleviate than the conventional traffic control problems with which the traffic engineer is usually confronted. The increasing severity of the problems is highlighted by the following two facts :

- On the one hand, congestion involves a waste of energy, an increase in traffic delays and a degradation of the environment by creating long queues of waiting vehicles. In congested conditions these effects are many times more severe than those occurring in non-saturated conditions.
- On the other hand, the difficulty is that, as shown in the preceding Chapter, there is not yet an adequate basis for developing control actions that fully respond to the problems posed. More research is needed to ensure the applicability of the mathematical solutions under real-life traffic conditions. The information that can be derived from the experience of those who endeavoured to find a practical solution is scarce. Close co-operation between the leading research institutions and demonstration projects should lead to valuable developments in this field.

The present Chapter contains a critical analysis of the strategies and practical experience regarding congestion control.

IV.1 LOCAL INTERSECTIONS

Mathematical studies show that the solutions for an optimal control ask for a sufficient buffer space to accomodate queueing traffic. In all these cases a "bang-bang" method is used, i.e. the traffic control cycle has an extreme green-split, favouring any one of the phases at the expense of the others. This makes the application of such control strategies rather unrealistic.

However, when congestion persists only for a short period, another strategy can be used. One approach based on a more pragmatic assumption is that the optimisation strategy has to ensure that queue lengths or waiting times never exceed a pre-set limit. Again, this strategy will lead to a "bang-bang" like control, meaning that the traffic control cycle will at all times show a split with one phase having a very long green period and all others with a minimum green period.

Another theoretical possibility is the on-line queue management approach where, on the basis of actual traffic data, the cycles are adapted in such a way that the ratio of the queues on the arms of the intersection is kept constant. This requires on-line measurements of queues and arrivals.

When comparing congestion control with non-saturated control, it can be seen that both the parameters that should be optimised and the strategies differ very much. This is also the case for the empirically derived strategies, as described in the NCHRP Report No.194

where simple rules to make the best use of the available buffer space are given. Again, one sees that just like in the case of the mathematical solutions the strategies for congestion control differ greatly from those for non-saturated control. This means that a new approach is necessary when facing congestion problems.

IV.2 NETWORKS AND ARTERIALS

The first point to make is that in the case of saturated networks and arterials one should try to improve the performance of the critical intersection. One should then identify the appropriate storage or buffer links to store the excess demand. One may use all the available storage capacity or just select a few links which have a lot of capacity and which do not interfere with other traffic and which can be by-passed by public transport. Network gating holding the vehicles at the periphery would be the next step.

As shown previously, off-line models can be used successfully to set up a scheme for this purpose. However, it means a complete traffic management strategy, not a simple rearrangement of traffic lights.

IV.3 THE ROLE OF TRAFFIC CONTROL

The present state of knowledge and the measures currently taken in the case of near-saturation or saturation have been reviewed in the preceding Chapter. It is, in fact, extremely difficult to compare different strategies and to evaluate the role of control, given the lack of a common reference system for simulated or field tests.

In the following discussion, an attempt is made to suggest an approach towards a common reference system and to provide a few indications on how traffic control in periods of saturation may be assessed.

Chapter II defines saturation as a state in which any further deterioration, even slight, in the characteristics and/or any increase in traffic load brings about a sharp decline in the level of service. Figure 1 (Chapter II) demonstrates the irreversibility of the phenomenon while Figures 2 and 3 provide an illustration of the two types of measures that may be envisaged in treating saturation :

- preventative measures, and
- curative measures.

A theoretical evaluation of what is involved in these two types of measures may be easily made on the basis of graphs 1, 2 and 3 in Chapter II. It may be seen that :

- the contribution of preventative measures is a function of the difference in area between curves 1(2) and curve 1.2.3.4.5.6. of Figure 1 ;
- the contribution of curative measures is a function of the difference in area between curves (2) and (1) of Figure 3, before and after saturation.

IV.3.1 Qualitative Approach

An analysis of the relative roles of the above measures shows that qualitatively :

- the relative contribution, in terms of "quality of service", of the preventative measures is always greater than that of the curative measures. In all cases it is better to avoid saturation by expanding capacity than by treating the situation curatively ;
- the treatment of saturation by curative measures will necessarily entail a loss in quality of service before the point of saturation (point 2, Figure 3). This implies predicting variations in quality of service as a function of demand and time. This forecast should be sufficiently precise so as :

i) to avoid unnecessary cutbacks in the service offered, particularly if demand is not to grow to such a point that it leads to saturation ; or

ii) to ensure that the curative measures are instituted before demand has led to saturation.

- Furthermore, the part played by saturation treatment cannot be limited to the sole aspect of "quality of service". Thus, in the case of a "successful" treatment of saturation - whether by preventative or curative measures - at any given level of demand there is a corresponding quality of service. If this analysis of level of service is effected by mode it becomes possible to organise journeys in such a way as to guarantee - by control and infrastructural measures - a given quality of service for a given demand. This point is of particular importance in the case of the "public transport" service level when the aim is to harmonise control and other actions (parking, management, tariff structure, etc.), designed to promote a modal transfer to public transport.

IV.3.2 Quantitative Approach

The quantitative approach to the treatment of saturation requires that it is at all times possible to draw the "quality of service-demand" curve, as well as, for the sake of completeness, the demand curve as a function of time.

Quality of Service

The difficulties in selecting the types of parameter qualifying the service levels and transcribing them are, moreover, compounded by the problem of selecting the period for which measurement data are aggregated as it is impossible to speak of any instantaneous value of a quality of service. Thus, according to the period of integration of the measurements taken, saturation may or may not be revealed. The problem is particularly marked in the case of some towns where the midday peak leads to queues and hence saturation over a very short time. If the data are integrated over periods longer than the length of the phenomenon, it could be said that - in the terms of the definition proposed - saturation has not occurred.

Demand

The measurement of demand poses the following difficulty : demand can be measured only in a free flowing situation, before it has been skimmed off by supply (otherwise it is supply rather than demand which is measured). Furthermore, demand at one point must be considered as being made up of the users who would cross this point were the supply unlimited (upstream demand increased by the number of vehicles queueing). Measuring this demand presupposes a knowledge of what the situation would be if demand were not limited by supply.

It is necessary to know how demand changes over time in order to make a quantitative assessment of the part played by the various factors for : on the one hand, with constant demand exceeding supply - at at least one point - the quality of service deteriorates with time (the queues lengthen) ; on the other, the "quality of service-demand" curve is not enough in itself to quantify the benefits to be derived from the control.

IV.4 CONCLUSIONS

Chapter III shows that the measures currently available and based on theoretical models tend to be of the preventative type, even when they are applied after saturation has occurred. This seems strange but in fact all these strategies aim at handling the given situation. No strategy prevents saturation through measures based on prediction.

Finally, as far as parameters are concerned, the standard parameter queue length is a very good one to detect saturation. It can be of help in curative types of strategies. But for preventative and curative strategies the main input data will be inflow/outflow information. The curative measures currently applied are essentially empirical.

Chapter III nevertheless points to certain actions which may be taken both in the case of near saturation and saturation.

The implementation of measures to enhance their roles outlined above calls for a better approach to what demand consists of and the time scales to be employed, as well as for more reliable methods and algorithms than those currently available, to enable the resulting demand and service levels to be predicted in time and space.

V

CONSIDERING FUTURE WORK

V.1 DEFINITIONS AND MEASURES

Congestion can be easily perceived but there have been only a few attempts to define and measure congestion. Most measurements have been taken as part of before-and-after studies for particular schemes and are aimed at measuring improvements achieved. Traffic characteristics, such as speed and flow, of some cities have been charted over several years and are sometimes referred to as the measure of congestion in these cities. While they give information about relative congestion from year to year they are not direct measures of congestion. Though speed is affected by congestion in the network, it does not indicate the underlying reasons for it. Since congestion in saturated conditions may be dealt with by control algorithms that are different from those for congestion in undersaturated or near saturated conditions, it is important that qualitative as well as quantitative measures are made, leading to conclusions as to the type of control system needed. In Chapter II several basic types of congestion are defined but these are often components of congestion from which a city will normally suffer. How the relative importance of the components should be decided and which components best characterise congestion needs further work.

For undersaturated traffic conditions the measures used generally are volume, occupancy and speed or journey time. The timescale or "horizon" over which these values are measured is not critical for undersaturated conditions, however in saturated conditions the timescale over which a parameter such as demand persists can be critical. For example it is meaningless to give a value of 110 % as a level of saturation without clearly stating the length of time for which the problem persists.

Another dimension of congestion is that of area of extent. Phrases such as "widespread congestion" say little about the problem and how it should be tackled. Smeed reported on the travel characteristics and expected journey speeds related to area and type of city. This may be a useful approach for further research.

When considering what information to collect three types of data can be specified :

a. Identification of congestion

b. Control of congestion

c. Effectiveness of control.

V.1.1. Measures for Identification of Congestion

Research is needed on how best to categorise the type and severity of congestion. In Chapter II the existence of a hysteresis effect in the demand/level of service curve was taken as typical for saturation. Measures which show up this hysteresis phenomenon will therefore be the most convincing. At a single junction, data on journey time and demand over a period will show up both the hysteresis effect and the time correlation of measures such as demand and level of service. The need to collect data in such a way that the hysteresis phenomenon stands out may lead to a change in data collection procedures so as to avoid too many averaging operations.

The use of journey time/flow or demand/service curves on an arterial or a network may give useful indications on the scale of congestion but will not necessarily point to the cause. To identify the cause of saturated conditions it is likely that queueing phenomena must be measured.

A crucial form of congestion which the chosen parameter(s) should identify is the blocking of upstream junctions by queues. This may be the major cause of congested conditions in city centres, although the interaction of junction blocking and lack of capacity is so complex that it is often difficult to determine which factor is the cause and therefore the target for control.

CONCLUSION : it is obvious that the definition of a useful parameter to identify congestion is needed.

V.1.2 Measures for Control

There are two important types of control measures: strategic measures which may be used to select a control policy or regime and tactical measures which will be used by these control algorithms.

The most common form of congestion is that which happens each day during the peak in most cities and the best strategic information may well be the use of historical average flow patterns. Data from Toronto used by Pignataro *et al* shows the high repeatability of traffic patterns that typifies many cities. Plan selection systems often use a combination of flow and occupancy values such that plans are changed when certain threshold values are reached. The Toronto data also shows that attainment of given flow levels occurs with great regularity of time of day and the debate between the selection of plans by time of day or by flow remains unresolved.

There are many models for continuous prediction of flows (ASCOT, Toronto, UTCS) and some for queue length (Barras & Levine) for undersaturated conditions, but Barras & Levine point out that the prediction performance is degraded by high levels of saturation. Prediction of both strategic and tactical parameters under saturated conditions is a major research area.

In undersaturated control, errors in detectors or predictors for parameters such as queue length are not likely to be cumulative because queues clear each cycle. However, under saturated conditions, the same predictors may be rendered useless by cumulative errors. Unfortunately the errors in any predicted value are likely to be inversely related to speed of response of the control strategy which is also important. The trade-off between accuracy and speed of response is often a critical factor in control systems.

One problem that arises with the transition from free flow to saturated flow is the two-valued nature of the speed flow curve. Occupancy can be used, but for very short-term measurement the variability in vehicle length creates problems. Research into measurement of control parameters throughout the range of traffic conditions and perhaps into the use of new parameters is needed. Sims and Dobinson mention the use in Sydney of "space time" which is the time a stopline loop is not occupied. This parameter avoids the disadvantage of vehicle length and by choice of loop length can be single-valued with speed.

There are a wide range of parameters associated with queues. Pignataro *et al* suggest :

1. Queue length
2. Link length minus queue length
3. Ratio of queue length to link length

Link length minus queue length is suggested as their preferred measure of the effects of saturation. Although congestion is not possible without queues, quantitative measurements of queues is difficult because,like congestion itself, queues are not precisely defined and not easily amenable to automatic measurement. Queue detectors may be useful but as their name implies, they detect that a queue exists at a point and are not queue measurement devices. Headway or occupancy detectors give more information about queueing conditions immediately upstream and downstream of the detector but then the position of detectors is crucial for useful information. If needed long loops can also give queue information.

The FHWA are funding a project at the Jet Propulsion Laboratory investigating the effects of Closed Circuit Television for data collection. The Wide Area Detector System (SCAN) is a new detection system which will be used to implement future control strategies. It utilizes a TV camera or other imaging detector, such as an infra-red imaging sensor, to scan a section of road or an approach to a signalised intersection and to track all vehicle movements within it. This approach has two advantages over current point detection systems :

1. relatively low installation cost,
2. the ability to measure spatial traffic flow parameters such as queues.

SCAN's development is currently centered on highway applications where it is being evaluated as a tool for measuring the average vehicle flow parameters - volume, density, and space mean speed. Future work on the SCAN may include evaluating its application to traffic signal control where it can be used for queue estimation as well as the usual measurements now obtained from point detectors.

Any new devices for measurement of control parameters must be considered not only for the information that they will theoretically provide, but also for their ease of maintenance, the time and level of compêtence needed to set them up and keep them in tune and their practical reliability. Indeed experience has shown that new devices need to be operated at at least 20 sites for periods of over a year before valid conclusions can be reached on their general utility.

CONCLUSION : In/out flow and queue are the best control parameters. The former can be measured easily,the latter is still more difficult to measure.

V.1.3 Measures of Effectiveness

Journey time or delay have long been considered the major measures of effectiveness of a traffic control system and are still the predominant measures. However, other measures are needed to fully describe the effectiveness of a saturated control system. These will usually be queue-based measures as the aim of many policies will be to restrict queues on certain links or in certain areas. There is a wide choice of measures even within the confines of journey time and queues. Kay gives the following from a report by Cooper et al.

Journey time	Queue
a. total in system	a. total in system
b. mean in system	b. mean in system
c. mean individual through system	c. mean at worst intersection
d. maximum individual through system	d. maximum at worst intersection.
e. mean on slowest link	
f. maximum on slowest link.	

Which type of measure will be the best to use is not certain. It may depend on the problem or simply on ease of measurement, but ought not to depend on the type of strategy.

If the measure of effectiveness is an input of the strategy then this makes comparison between different strategies installed in different cities difficult if not impossible.

If strategies are to be compared then a baseline system is needed. What this might be is not clear and would merit research in itself. Such a baseline system must be relatively cheap to buy and install.

The time-of-day fixed time system which has become a standard for free flow conditions may be tuned to oversaturation as when testing SATURNE. This tuned fixed plan may serve as a baseline although what tuning should be done and how is not all clear and should be an area of further work.

CONCLUSION : Effectiveness can not yet be expressed in a commonly accepted formula like TRANSYT's P.I. It will be some combination of queue and journey time. In practice effectiveness will probably be expressed in terms of presence or absence of spillback, or in terms of general area mobility.

V.2 ISOLATED INTERSECTIONS

Much of the basic research into saturated conditions has concentrated on the isolated intersection because it is a simpler problem than the network's. The research has in the main considered saturated conditions where the junction demand is greater than the junction capacity. However, a large number of practical systems include saturation caused by short-term high demand and random cyclic flows. The simplifying assumption of queues on all arms is often not valid ; different arms of a junction my be oversaturated at different times with total junction demand fluctuating near capacity for much of the peak period with short periods of true saturation. The tendency of drivers to avoid saturated junctions may ensure that near saturated conditions are common at many junctions throughout a network. The problems caused by this type of situation need solving with similar urgency to the relatively straightforward fully saturated junction problem.

Among modern approaches to vehicle actuated control in free flow conditions the work by Miller has proved to be of great interest. Work in Sweden by Bång and in France by Amy de la Breteque show both in simulation and on-street that it is an effective method. Extensions to the saturated case by this and other methods suggested by Langley, Church, Gordon, Michalopolous and others need testing to provide information about their effectiveness of control over a range of flows from near saturation to saturation. Whether any one strategy will be sufficient to cover all conditions is uncertain and a combination of methods may be needed with smooth transition between methods as an essential element.

The lack of implementation of research ideas is of concern. It may be partly due to the practical difficulty of modifying present controllers to the demands of new and sometimes complex strategies, for example there may be a case for a complete reconfiguration of stages or phase groups to suit different conditions. Often, however, research is aimed at solving the problems which are analytically tractable rather than solving the problems that exist on street. The gap between research and practice remains large and it is this gap which must be filled if effective control schemes are to be produced. While it is not possible for many researchers to implement and measure their ideas on street it is possible for them to consider such areas as error sensitivity, default modes for faulty equipment, robustness of logic to the random events which happen every day on street such as parking, loading, roadworks and accidents.

Some free flow strategies presently in use point to difficulties which should not be repeated in saturation strategies. Present controllers can be difficult for the local traffic engineer to set up and fine tuning is not always possible. Furthermore it is sometimes difficult to know whether the controller, when set up, is working correctly. These difficulties may be compounded by more complex controllers unless careful thought is given to both the level of competence needed to set up the controller and the amount of time required for setting up and maintenance. This is especially important for saturated control because the consequences of a poorly set-up and poorly maintained system can be much worse in saturated conditions than under free flow. One answer to the difficulties of setting up controllers is to use automatic procedures for as many parameters as possible. An example is the automatic calculation of saturation flow as it varies both throughout the greentime and throughout the day.

The key to success or failure may be in the effective use of the microprocessor in local controllers. The extra intelligence is already used for safety constraints, implementation of timings and fault checking in commercially available controllers, but the use of the microprocessor to implement novel strategies is rare at present. The intelligent controller must be geared to the needs of the practical traffic engineer if the efficiency of control is not to be strictly limited. The use of microprocessors may widen the gap between product and user unless great care is taken to "human engineer" the product, to use the phrase of an American company.

CONCLUSION : Research should be aimed at practically implementable strategies which are robust, not at mathematical sophistication.

V.3 CONTROL MODES FOR NETWORKS

Implementations of true congestion control are rare, with Bordeaux as the main example. There have been several attempts at queue management by traffic signals. Some such as in Bern and Southampton were successful, others such as the Nottingham experiment, were not. It should be noted that none of the last three were congestion control only but had, as their main aim, improvements in public transport. In Bern and Southampton reductions in bus journey times were sought at the cost of other vehicles. In Nottingham the intention was to significantly increase delay to private vehicles in order to cause a shift in ridership from cars to public transport. There is as much to learn from the failure in Nottingham as from the success elsewhere. The reasons for failure were often in the details rather than the initial concept.

The schemes mentioned reinforce the point that the objectives and aims of a control system must be quite clear and the limits of applicability of a system must be considered. They also emphasise the difficulty in producing a standard system applicable to all cities and situations.

From the few congestion control systems that do exist and from free flow systems, it is apparent that there are three categories of congestion control, each of which will be attractive and suitable for different cities and types of congestion. These are :

1. Fixed Time Plans. This category will require work on the development of congestion plans and the implementation of these plans to best effect.
2. Fixed Time Plan Background with Local Adaptation. In free flow traffic control this area has been difficult to justify in cost effectiveness over a normal fixed time system due to the need for both detection and for development of fixed time plans. However, for congestion control the extra degree of flexibility allowed by some real-time adaptation may be necessary for successful operation of fixed time systems.
3. Fully Adaptive Systems. It is only fairly recently that fully adaptive free flow systems have been able to measure up to the success of fixed time strategies. The same process of development from fixed time to adaptive systems may be necessary for congestion control systems although the experience gained in fully adaptive free flow systems may allow parallel development of adaptive and fixed time control.

Each of these types may have sub-categories,the most obvious of which is the relationship of the detection and the control to the site of the congestion problem. The Bitterne Road system for example, gates traffic up to 2 km away in order to avoid congestion at a bottleneck river bridge.

V.3.1 Fixed Time Control

In many cities congestion is a regular problem during the peak hours of each day. Whether the congestion is at a few critical junctions or is a network-wide problem, the time and space distribution of the congestion is often known. In this situation it is possible that a fixed time plan can be used to limit the arrivals at critical points to avoid saturation or, if unavoidable, to reduce its worse effects.

The offline optimisation programs for free flow fixed time plans (TRANSYT, SIGOP, THESEE) need a lot of traffic information such as flow, turning movements, etc., but the data is reasonably easy to collect. For the saturated case extra data will be needed for any offline program and much of it will require more traffic engineering judgment such as designation of storage links and relative importance of links in terms of delay and acceptable queue length.

Many optimisation techniques have been suggested (Saturne, Tödli, Langley) for offline signal plans and further investigation is needed to consider both the users problem in setting up such programs and the stability and applicability to a range of real networks.

The need for origin-destination information in some methods (d'Ans, Tough) has the disadvantage that such data is difficult and costly to collect. Methods are being researched in the United Kingdom (Willumsen), the United States and France to produce O-D information from flow data. Although the result is not unique the information may well be good enough to use in optimisation methods.

There is still a gap between engineering solutions and research methods such as linear programming. An example is the concept of soft constraints. This implies not a rigid limit for a value such as queue length, but an increasingly severe penalty for exceeding a threshold value. In a saturated network, though it is not desirable to block major junctions, the best traffic solution might involve the blocking of several minor junctions. The simple approach of either no queue limit or a rigid limit may not be sufficiently flexible. Piecewise linear programming or even non-linear programming may be necessary to produce solutions which are acceptable on-street.

Linear programming type methods may be used to impose a structure on a network in terms of reservoirs for queueing or relative capacities with a detailed hill-climbing optimisation being used to produce the final FT plan. Such a hybrid of linear programming and hill climbing may be a type of solution worth pursuing.

CONCLUSION : It is likely that fixed time plans will be used in conjunction with information detectors to switch from plan to plan so that the control employed matches the demand and also avoids blocking critical junctions. If only one plan is used then all storage reservoirs must be filled together to ensure that those near to the critical points do not overfill and cause congestion themselves. However, restraint, if applied, should be applied close to the critical points if excess unnecessary delay is not to be caused to vehicles on unrelated routes. More than one plan allows sequential filling of reservoirs but imposes plan change transients. The number of plans and timing of the plan changes need careful study, as does the way to change plans. In summary, these methods are implementable, but improved access to obtain details of these methods should be ensured.

V.3.2 Fixed Time Control with Local Adaption.

Much of the theory of control of saturated conditions can be allocated to either the fixed time or the fully adaptive area. However, the use of local adaptation in addition to a background fixed time plan may often be the desired practical solution. It is often possible to identify critical junctions and areas where local adaptation will improve conditions with minimum interaction throughout the area. The need is mainly for simulation and implementation experience rather than new algorithms.

The approach being used in THESEE SATURATION (Texier) exemplifies this type of control. It has the advantage of response to saturation as it appears but is heavily constrained by the background plan and hence relatively stable. Experiments in Liverpool (Davison) point to this type of solution though no general conclusion about its effectiveness throughout a network can be drawn.

CONCLUSION : What has been concluded above about fixed time plans is also valid in this case. Local adaptation is a desirable option, but not more.

V.3.3 Fully Adaptive Control

Fully adaptive control implies that the signal timings are updated by the control system and no form of present plan is used. There is a wide range of types of adaptive control from an unconstrained system without knowledge of time of day to a more limited system which uses a lot of time of day information to supplement the detector information. At present the approach to fully adaptive systems has been twofold :

1. Data-oriented systems (empirical, specific)
   This system is often favoured by city traffic engineers and is often initiated by them. The starting point is a large amount of local knowledge and a traffic engineering insight into practical solutions. This approach is exemplified by the GERTRUDE system in Bordeaux and the SCAT system in Sydney. The aim is often to achieve maximum flexibility but it does require a skilled traffic engineer to set up the data which may be extensive and esoteric.
   These systems are at present working and control congestion in their designed locations. Their portability to other systems relies as much on a transfer of traffic engineering skill and insight as in software or theory.

2. Control Algorithm-oriented systems (theoretical, general)
   When the problem of congestion control is investigated from a research point of view, the first aim is often to understand and perhaps to model the congestion and this leads to the development of an algorithm to control one or more junctions. Information on a given application such as street layout, choice of reservoirs and detector information processing is secondary. This can mean that designs are elegant but impractical, though when successful they are not site-dependent and may be easily portable from one city to another. These methods may be most useful in cities that lack large amounts of traffic engineering support.

A fruitful area for research might well be in the hybrid of these two types which, while allowing flexibility to a skilled traffic engineer to improve, adjust and add his intuitive insight into his city's problems, could nevertheless, if necessary, be set up by standard procedures and with limited intuitive input from the traffic engineers.

In free flow control, fully adaptive systems are being developed of both types and the hybrid is an attainable goal as shown in systems such as the SCOOT system. Extending the control to saturated systems may initially be easier by the empirical approach. This will require extra data often of an, at present, undefined type such as the undesirability of queueing on a link, reservoir hierarchies and alternative assigned routes. However, it is important that the theoretical problem is solved on the street to allow the development of both easily portable systems and the desired hybrid of a theoretical and empirical system.

One of the major problems of free flow fully adaptive systems has been the trade-off between computing time for prediction and signal calculations and the required response time of an adaptive system. This problem may well be worse for saturated systems in that the consequences of delayed implementation can be more severe and longlasting. Whether hill-climbing or linear programming will have adequate complexity to avoid oversimple assumptions and at the same time be fast enough for implementation in real time is not yet certain. The decision of the Metropolitan Toronto traffic authorities to buy an array processor for on-line optimisation for free flow conditions points to the possibility of greatly increased computer power in the future for traffic signal optimisation.

CONCLUSION : It is clear that research should be aimed at a hybrid model for network control. It may well take many years to obtain valid results. Due to the excess demand problems it is expected that it will have to contain some kind of O-D matrix (on-line, updated) or based on gating procedures.

V.3.4 Simulation and On-line Models

When developing control strategies for oversaturated networks the use of simulation is an essential prerequisite to on-street implementation. Simulation of saturated conditions is especially difficult because of the complex behaviour of drivers within a junction that is partially blocked by a queue. The detail of a NETSIM-type model will be useful for researching the problems of oversaturation but may be too large in computer resources for many research projects. Simple, crude, special-purpose simulation may allow effective research with greatly reduced computing costs. Development of special-purpose simulators for congested conditions will be of great help in the testing of control algorithms.

At present most adaptive strategies rely on algorithms which relate street information such as flow and occupancy directly with control parameters such as green-time (or change of green-time) offset and cycle time. As computers become more powerful it has become possible to use an on-line model of traffic conditions. The on-line model is an intermediate function between detector processing, prediction etc. and signal control decisions. In the SCOOT adaptive control system designed for below or near saturation the effects of vehicles and platoons are modelled on each link. The on-line model gives information on queues, flows, delays, stops and percentage saturation on each link.

The information from an on-line traffic model can be used in two ways. First it can be used by optimising algorithms to produce better signal timings, and second it gives the traffic engineer up-to-date knowledge of his network. This information can confirm the efficacy of the system in a way which is more difficult with a non-behavioural approach. The data can also be used to gain a greater understanding of the area at a traffic management level. Colour graphics is an extremely useful medium for information presentation.

CONCLUSION : There is still much more to be done with on-line models, for example determining what is the best input information - flow and/or occupancy -, accuracy and error sensitivity, update speeds and data presentation, and taking into account the time varying effects.

V.4 PUBLIC TRANSPORT VEHICLES IN CONGESTED CONDITIONS

This is a major sector for further research both empirical and theoretical. A review of the problems and of some of the solutions actually implemented in Member countries is contained in Section III.8. There is an obvious need for systematic research in this field.

V.5 BEHAVIOURAL EFFECTS

Although the main thrust of research will be into technical solutions to the problem of congestion, there are other areas of research which will be important in finding strategies which are effective.

The perception of congestion by drivers and their willingness to re-route or to re-time their journeys is an area where there is little hard knowledge to influence traffic system design.

It is likely that assignment techniques will be too complex and time consuming for use in congestion control. However, the need to investigate the effects of the control strategy on route choice and diversion routes will necessitate the use of detailed traffic assignment models. Many optimal control policies leading to "bang-bang" solutions can lead to an inequitable solution for many drivers. An alternative strategy which minimises individual delay gives a maximum individual delay to drivers. Driver reactions to inequity in delay is important : for example, to what extent will drivers tolerate higher delays on side roads than main roads and should this tolerance of inequity be used to improve overall efficiency ? Similarly the acceptability of long delays in queue reservoirs as against being in a slow-moving continuous queue may determine which strategies will gain acceptance with policy makers, traffic engineers and the public at large.

BIBLIOGRAPHIE

General

ATEC, Second International ATEC Congress, "Regulation 80", Paris, April 1980.

CETUR, "Traitement de la saturation - Approche théorique et applications pratiques", Paris, March 1980.

IFAC-IFIP, First International Symposium on Traffic Control, Versailles, June 1970.

IFAC-IFIP, Second International Symposium on Traffic Control and Transportation Systems, Monte Carlo, September 1974.

ITS, U.S. IHWA, Proceedings of the International Symposium on Traffic Control Systems, Berkeley, August-December 1979.

NATO, CCMS, "Bus Priority Systems", Report No. 45, 1976.

OECD, Road Research, "Electronic Aids for Freeway Operation", OECD, Paris 1971.

OECD, Road Research, "Area Traffic Control Systems", OECD, Paris 1972.

OECD, Road Research, "Traffic Operation at Sites of Temporary Obstruction", OECD, Paris 1973.

OECD, Road Research, "Capacity of At-grade Junctions", OECD, Paris 1974.

OECD, Road Research, "Research on Traffic Corridor Control", OECD, Paris 1975.

OECD, Road Research, "Bus Lanes and Busway Systems", OECD, Paris 1977.

OECD, Road Research, "Integrated Urban Traffic Management", OECD, Paris 1978.

OECD, Road Research, "Traffic Measurement Methods for Urban and Suburban Areas", OECD, Paris, 1979.

OTA, 8. Intern. Studienwoche für Strassenberkehrstechnik, Thema V : Verkehrsregelung v. städtischen Strassennetzen, Barcelona, Sept. 1966.

Alphabetical

AKCELIK,R., MAHER,M.J., "Route Control of Traffic in Urban Road Networks", Transpn. Res. Vol. II, pp. 15-24, Pergamon Press 1977.

ALBRECHT, H., PHILIPPS, P., "Ein Programmsystem zur verkehrsabhängigen Signalsteuerung nach dem Verfahren der Signalprogrammbildung (PBiL)", Heft 240, 1977.

ALLSOP,R.E., "SIGSET - A Computer Program for Calculating Traffic Signal Setting", Traffic Engineering & Control, June 1971.

ALLSOP,R.E., "SIGCAP - A Computer Program for Assessing the Traffic Capacity of Signal Controlled Road Junctions", Traffic Engineering & Control, August 1976.

AMY de la BRETEQUE et al, "Adaptive Control at an Isolated Intersection T.E. & C", July 1979.

ANCKER, J.C. et al, "The Oversaturated Intersection, Some Statistics", TM-3478/002/00, Syst. Dev. Corp., Santa Monica, Cal., United States.

ANCKER, J.C. and GAFARIAN, A.V., "A Simple Renewal Model of Throughput at an Oversaturated Signalised Intersection", Transportation Research, Vol. 1, No. 1, May 1967.

ANON, "Gertrude, Bordeaux - A. Bordeaux : une stratégie originale de régulation du trafic". Le Moniteur des travaux publics et du Bâtiment, p. 87, 13th March 1978.

D'ANS, G.C. and GAZIS, D.C., "Optimal Control of Oversaturated Store-and Forward", Transportation Networks, Transport SCI, Vol. 101, pp. 1-19, Febr. 1976.

BACON,W., "Linked Traffic Signals for Maximum Capacity", Transp. Res., Vol. 11, pp. 229-34.

BACON,W., "The Role of Signals in Traffic Management", The Highway Engineer, March 1975.

BARRAS,J.S., et al, "Discrete Time Point Processes in Urban Traffic Queue Estimation", I.E.E.E. Transactions, AC 24, No. 1, 1979.

BENNET,B.T. and GAZIS,D.C., "Control of the Lincoln Tunnel Traffic by an On-line Digital Computer", Fourth International Symposium on Theory of Traffic Flow, Karlsruhe, 1968.

BIROS,J.F., "La Régulation des Feux Tricolores en Milieu Urbain", Revue Générale des Routes et des Aérodromes, March 1977.

BOESEFELDT,J., EVERTS,K., PHILIPPS,P., "Untersuchungen zur verkehrsabhängigen Signalsteuerung - Teil B : Steuerungsmodelle für Strassenzüge", Strassenbau und Strassenverkehrstechnik, Heft 131, 1972 ; "Teil C : Steuerungsmodelle für Teilnetze", Heft 149, 1973 ; and "Teil D : Integrierte Betriebssysteme für Netze", Heft 188, 1975.

BOESEFELDT,J., and EVERTS,K.,"Traffic Guidance and Information Systems for Motorway Networks. - Strategy and Application", Int. Symp. on Traffic Control Systems, Berkeley 1979.

BOTMA,H., "A Scheme for Evaluating a Local Queue Warning System", Traffic Engineering & Control, Vol. 19, No. 7, p. 350, July 1978.

BRANSTON, D., "A Comparison of Observed and Estimated Queue Lengths at Oversaturated Traffic Signals", Traffic Engineering & Control, Vol. 19, No. 7, p. 322, July 1978.

BRETHERTON, R.D., "Five Methods of Changing Fixed Time Traffic Signal Plans", TRRL report, LR 879.

BURHART, K., "Urban Traffic System Optimisation", Univ. of Minnesota, 1971.

BÅNG, K.L., "Optimal Control of Isolated Traffic Signals", Traffic Engineering and Control, July 1976.

CANADIAN MINISTRY OF TRANSPORT, "Improved Operation of Urban Transportation Systems", March 1976, Vol. 1 - "Traffic Signal Control Strategies - A State of the Art", Nov. 1975, Vol. 2 - "Evaluation of Off-line Area Traffic Control Strategies", Nov. 1976, Vol. 3 - "Development and Evaluation of a Real Time Computerised Traffic Control Strategy", Ottawa, 1975/76.

CANADIAN MINISTRY OF TRANSP. & COMM., "Q.E.W. Freeway Surveillance and Traffic Control Demonstration Project", June 1974.

CHANG, A., "Synchronization of Traffic Signals in Grid Networks", I.B.M. Journal of Research & Development, New York, July 1967.

CHU, K.C., "Decentralised Real-time Control of Congested Traffic Networks", Proceedings of the Seventh Intern. Symposium on Transportation and Traffic Theory, pp. 471-491, Kyoto, 1977.

CHURCH, R., and REVELLE, C., "Modelling an Oversaturated Intersection", Transportation Research, Vol. 12, No. 3, June 1978.

CILLIERS, M.P., et al, "FREQ6PL - A Freeway Priority Lane Simulation Model", Institute of Transportation Studies, University of California, Report No. UCB-ITS-RR-78-8, Berkeley, California 94720, August 1978. California 94720, August 1978.

COOPER, D.L., and WALINCHUS, R.J., "Measures of Effectiveness for Urban Traffic Control Systems", Highway Research Bord Record, pp. 46-59, Washington, 1970.

COOPER, D.L. et al, "Final Report, System Analysis Methodology in Urban Traffic Control Systems", F.R.W. Houston Texas, June 1979.

CULSHAW, T.A., MORRIS, R.W.J. and PAK-POY, P.G., "Comparison of Traffic Signal Controllers Using Computer Simulation - Some Preliminary Results", A.R.R.B., 3:1, Melbourne 1966.

DAVID, Y., and HENRY, J.J., "Saturne ou la Régulation des Feux en Période de Saturation", Transport, Environnement, Circulation, No. 17, Paris Juillet/Août 1976.

DAVISON, A., and THORPE, R., "Reviving Circulation on our Arteries", The Surveyor, 14th September 1978.

DAWSON, R.F.F., and VASS, P., "Vehicle Operating Costs", TRRL, Lab. Report 661.

DEBARTHE, S., "Reconstitution de Matrices Origines-Destination en Milieu Urbain", TEC., No. 34/35, Paris 1979.

DEPARTMENT OF THE ENVIRONMENT, "Bus Demonstration Project - Summary Report No. 8 Southampton. A Linked System of Traffic Signals with Priority Access to the Route for Buses", London 1976.

DIRECTION REGIONALE DE L'EQUIPEMENT D'ILE DE FRANCE, "Opération "Matin d'Ile de France" - premier bilan".

DORSEY, W.F., "Highway Advisory Radio Operational Site Survey and Broadcast Equipment Guide", U.S. Department of Commerce, National Technical Information Service, 5285 Port Royal Road, Springfield, Virginia 22161, PB No. 301379, April 1979.

DUDEK, C.L., "Real-Time Displays", Proceedings of the International Symposium on Traffic Control Systems, Vol. 1 - Resource Papers, U.S. Department of Commerce, National Technical Information Service, 5285 Port Royal Road, Springfield, Virginia, PB No. 80-105091, August 1979.

DUNNE, M.C. & POTTS, R.B., "Algorithm for Traffic Control", Op. Res. No. 12, pp. 870-81, 1964.

DUPUY, A.C., KITTLESON, W.K., and MAY, A.D., "Evaluation of Selected Surface Street Network Models", Freeway Operations Study - Phase IV, Report No. 74-3, The Institute of Transportation and Traffic Engineering, Berkeley, California, June 1974.

EISNER, M.J., "Synchronizing Traffic Signals in a Saturated Network, a Deterministic Optimisation Model", Tech. Report No. 155, Cornell University.

ELMBERG, M., "Compartmentation as a Tool to Reduce Traffic Congestion and Improve the Environment", Special Report 153, US Transportation Research Board, Washington D.C., 1975.

EVERALL, P.F., "Urban Freeway Surveillance and Control, The State of the Art", Superintendent of Documents, US Government Printing Office, Washington, D.C. 20402, June 1973.

F.H.W.A., "Evaluation of UTCS Control Strategies", Report No. FHWA-RD-76-149, Washington, July 1976.

F.H.W.A., "Application of UTCS First Generation Control Software in New Orleans", Report No. FHWA-RD-78-2, Washington, January 1978.

FORSCHUNGSGESELLSHAFT FÜR DAS STRASSENWESEN, "Merkblatt für die Beschreibung von Vorbereitungs- und Steuerungsmodellen zur innerörtlichen Verkehrslenkung, "PBIL), Cologne 1977.

FOWKES, N.D., "Road Traffic Flow in Urban Centres, a New Approach", Transp. Research, Vol. 6, pp. 39-48.

FRANCERIES, C., "Un Centre-Ville sans Embouteillage", Revue Générale des Routes et des Aérodromes, March 1977.

FRANCERIES, C., "Circuler en Agglomération Bordelaise - Gertrude", Le Moniteur des Travaux Publics et du Bâtiment, No. 10, March 1978.

FUEHRER, H.H., "Signal Plan Generation : Present State and Future Developments", Inst. Elec. Engrs. Colloquium on Area Traffic Control, London, May 1972.

GARCIA, RAMON, J. et al, "New Area Traffic Control System in the City of Barcelona", Second Int. Symp. Traffic Control and Transportation Systems, IFAC-IFIP, Monte-Carlo, September 1974.

GARTNER, N.H., et al, "Optimisation of Traffic Signal Settings by Mixed-Integer Linear Programming", Transp. Science G. pp. 321-343, 1975.

GAZIS, D.C., "Optimum Control of a System of Oversaturated Intersection", Operations Research, Vol. 12, pp. 815-831, 1964.

GAZIS, D.C., "Modelling and Optimal Control of Congested Transportation System", IBM Research, RC 3977,

GAZIS,D.C., and D'ANS, G.C., "Optimal Control of Oversaturated Store-and-Forward Transportation Networks", Transportation Science, Vol. 10, No. 1, pp. 1-19, February 1976.

GAZIS,D.C., and POTTS, R.B., "The Oversaturated Intersection", IBM Research Paper, RC-929, April 22, 1963.

GAZIS, D.C., and POTTS, R.B., "The Oversaturated Intersection", Proceedings, Second International Symposium on Traffic Theory, London, England, June 1963, published OECD Paris.

GERLOUGH, D.L., and WAGNER, F.A., "Improved Criteria for Traffic Signals at Individual Intersections", NCHRP Report 32, Highway Research Board, Washington, D.C., 1967.

GORDON, R.L., "A Technique for the Control of Traffic at Critical Intersections", Trans. Science, Vol. 3, No. 4, 1969.

GUILLAUME et al, "Comparison of Three Urban Traffic Assignment Models", Report No. 29, IRT, Arcueil, Feb. 1978.

GUN, Y.J., et al, "Estimation of an O-D Trip Table Based on Observed Link Volumes and Turning Movements, Vol. I, US D.O.T. - FH-11 9292, Final Rep. 1979.

HALLWORTH, M.S.,"High Capacity Signal Design", T.E. 2.C., Vol. 21, No. 2, Feb. 1980.

HENRY, J.J., et al, "Etude de la Régulation du Trafic Urbain en Période de Saturation", Final Report, Centre d'Etudes et de Recherche de Toulouse, 1973.

HOLROYD, J. and OWENS, D., "Measuring the Effectiveness of Area Traffic Control Systems", Road Research Laboratory, Department of the Environment, RRL Report LR 420, 1971.

HUDDART, K.W., and CHANDLER, M.J.H., "Progress of Area Traffic Control in London", Traffic Engineering & Control, pp. 192-5, September 1971.

HUDDART, K.W., and REKAY, I.S.N., "Dynamic Control of Traffic Queues", Traffic Engineering & Control, May 1976.

INST. DE RECHERCHE DES TRANSPORTS,"Programme PERCEVAL - Notice d'utilisation"(SERES/IRT), Arcueil.

JAMES, R.L., "Identify your Most Critical Traffic Problems", Traffic Engineering & Control, October 1971.

JENEZON, J.H., "Berekening van de extra economische kosten en brandstofverbruik als gevolg van filevorming op rÿkswegen", Report Dienst Verkeerskunde, Rykswaterstaat, Holland, 1979.

JOVANIS, P.P., et al, "FREQ6PE - A Freeway Priority Entry Control Simulation Model", Institute of Transportation Studies, University of California, Report No. UCB-ITS-RR-78-9, Berkeley, California 94720, November 1978.

KAY, J.L., "Measures of Effectiveness", Vol. I, Proc., Int. Symp. on Traffic Control Systems, Berkeley, 1979.

KIMBER, R.M., and HOLLES, E.M., "Peak-Period Traffic Delays at Road Junctions and Other Bottlenecks", TERC, October 1978.

KOBAYASHI, F., "Comprehensive Automobile Traffic Control System", Japanese Contribution to OECD Group TS5, Paris, June 1979.

KOSHI, M., "A New Method of Traffic-Responsive Control of Traffic Signals", International Symposium on Traffic Control Systems, Berkeley, August 1979.

LE COCQ, J.P., "Commande par le Trafic des Feux d'un Carrefour", Ministère de l'Equipement et du Logement. Division Circulation-Exploitation, Bagneux.

LIEBERMAN, E.B., and WOO, J.L., "SIGOP-II : A New Computer Program for Calculating Optimal Signal Timing Patterns", Transportation Research Record No. 596, Transportation Research Board, 2101 Constitution Avenue, N.W., Washington, D.C. 20418, January 1976.

LONGLEY, D., "A Control Strategy for a Congested Computer-Controlled Traffic Network", Transportation Research, Vol. 2 No. 1, pp. 391-408, November 1968.

MacGOWAN, C.J., "Bus Priority and Bus Preemption in the Urban Traffic Control System", Proceedings of the 45th Annual Meeting of the Institute of Traffic Engineers, Institute of Transportation Engineers, 525 School Street, S.W., Washington, D.C. 20024, August 1975.

MacGOWAN, J., and FULLERTON, I.J., "UTCS - Development and Testing of Advanced Control Strategies in the Urban Traffic Control System", Public Roads magazine, Vol. 43, Nos. 2, 3 and 4, Superintendent of Documents, U.S. Government Printing Office, Washington, D.C. 20402, 1979-80.

MacLEOD C.J. et al, "Optimum Split of Signal Cycle Time for Road Junctions Within an Urban Network", Transp. Res. Vol. 11, pp. 183-8.

MARCONI, W., "The Relative Efficiency of Various Types of Traffic Controls", Traffic Engineering, April 1963.

MARLOW, M. and EVANS, R., "Urban Congestion - Survey 1976", TRRL Supplementary, Report 438

McSHANE, W.R., and CROWLEY, K.W., "The Regularity of Some Detector-Observed Arterial Traffic Volume Characteristics", presented before the Transportation Research Board Annual Meeting, Washington, D.C., January, 1976.

MAY, A.D., "The Role of Planning Models in Transportation System Management", Research Report UCB - ITS - RR - 78 - 11, Berkeley, Cal. 1978.

MICHALOPOULOS, P.G., and STEPHANOPOULOS, G., "Oversaturated Signal Systems with Queue Length Constraints - I. Single intersections. II. Systems of intersections", Transportation Research, Vol. 11, No. 6, pp. 413-421 and 423-428.

MICHALOPOULOS, P.G. and STEPHANOPOULOS, G., "Optimal Control of Oversaturated Intersections", TEC, Vol. 19, No. 5, May 1978.

MILLER, A.J., "A Computer Control System for Traffic Networks", Proc. Second International Symposium on the Theory of Road Traffic Flow, London 1963, Published OECD Paris 1965.

MOORE, S., "Three Traffic Signal Control Philosophies, applied to an Arterial Traffic System", Dep. of Motor Transport, New South Wales, 1976.

MORIN, J.M., "La Régulation des Corridors Autoroutiers Urbains par le Contrôle d'Accès" Rapport de recherche, IRT No. 26, Arcueil.

MUNT, P.W., "Traffic Monitoring Review for 1978", GLC RM 555, London, 1979.

NEWELL, G.E., "Traffic Signal Synchronisation for High Flows on a Two-way Street", Quarterly Applied Mathematics 21, pp. 315-325, 1964.

NIP, K.F., "A comparative Study of Traffic Control Schemes at an Isolated Intersection, Traffic Engineering and Control, pp. 264-269, June 1975.

OKUTANI, I., "Traffic Signals Synchronization for an Oversaturated Network", presented at the Seventh International Symposium on Traffic Theory, Tokyo 1977.

PAK-POY and Associates, "A Comparison of Road Traffic Signal Controllers Using Simulation Techniques", A.R.R. Report No. 10, February 1975.

PAQUIN, R., "Stratégies de Commande de Réseaux Urbains Saturés", 2nd IFAC/IFIP Symposium on Traffic Control and Transportation Systems, Monte-Carlo, September 1974.

PHILIPPS, P., "Vergleich verschiedener mikroskopischer Steuerungsverfahren zur Verkehrsabhängigen Signalsteuerung", Forschungsauftrag d. Bundesverkehrsministeriums, Bonn, July 1977.

PIGNATARO, J.L., CROWLEY, K.W., and LEE, B., "Oversaturation Terms and Measurements Seem Adequate After Field Validation", Traffic Engineering, June 1974.

PIGNATARO, et al, "Traffic Control in Oversaturated Street Networks", Final Report of NCHRP Project 3-18 (2), Washington, D.C., June 1975.

PIGNATARO, J.L., CROWLEY, K.W., and LEE, B., "Better Use of Signals Under Oversaturated Flows", Special Report 153, pp. 50-62, U.S. Transportation Research Board, Washington, D.C. 1975.

PIGNATARO, J.L., et al, "Traffic Control in Oversaturated Street Networks", National Cooperative Highway Research Program (NCHRP) Report No. 194, Transportation Research Board, 2101 Constitution Avenue, N.W., Washington, D.C. 20418, 1978.

ROBERTSON, D.I., "Fuel Savings", TRRL Report, L.R. 787.

ROBERTSON, D.I., and BRETHERTON, R.D., "Optimum Control of an Intersection for any known Sequence of Vehicle Arrivals", 2nd IFAC/IFIP Symposium on Traffic Control and Transportation Systems, Monte-Carlo, September 1974.

ROBERTSON, D.I., "TRANSYT : A Traffic Network Study Tool", TRRL lab. Report LR. 253.

ROSENBLOOM, D.S., "Peak-Period Traffic Congestion", PTRC Summer Meeting 1978 at University of Warwick, 10-13 July, 1978. PTRC Summer Meeting 1978 at University of Warwick, 10-13 July, 1978, Transportation 7 (1978) pp. 167-191.

ROSS, D.W., SANDYS, R.C., and SCHLAEFLI, J.L., "A Computer Control Scheme for Critical Intersection Control in an Urban Network", Transportation Science, Vol. 5, No. 2, May 1971.

ROSS, D.W. et al, "Improved Control Logic for Use with Computer Controlled Traffic", Interim report, NCHRP Project 3-18(1), H.R.B., Washington 1972.

ROTHERCOCK, C.A., and KEEFER, L.E.,"Measurement of Urban Traffic Congestion", US Highway Research Board Bulletin, No. 156, Washington 1957.

SERVICE REGIONAL DE L'EQUIPEMENT DE LA REGION PARISIENNE, "Autoroute du Nord - Signalisation variable", Rapport d'enquête, Paris.

SERVICE REGIONAL DE L'EQUIPEMENT DE LA REGION PARISIENNE, "Autoroute du Nord - Expérience de Régulation du Trafic", Paris.

SIMS, A.G., and DOBINSON, K.W., "SCAT-The Sydney Coordinated Adaptive Traffic System - Philosophy and Benefits", Vol. 2B, Proceedings Int. Symp. on Traffic Control Systems, Berkeley, 1979.

SMITH, M.J., PTRC (ed) Proceedings of Seminar F., pp. 113-124, Summer Annual Meeting Univ. Warwick, England, 1977.

STEIERWALD, G., BOESEFELDT, J., EVERTS, K., KEUDEL, W., SCHÖNHARTING, J., "Untersuchungen zur verkehrsabhängigen Signalsteuerung", Strassenbau und Strassenverkehrstechnik, Heft 100, 1970.

TACHE, P., "Le coût de la congestion", TEC No. 31, pp. 16-21, Paris 1978.

TARNOFF, P.J. and PARSONSON, P.S., "Guidelines for Selecting Traffic Signal Control at Individual Intersections", Draft Final Report for NCHRP Project 3-27, Washington, D.C., September 1978.

TEXIER, P.Y., "Rôle de la régulation dans le traitement de la saturation", TEC, No.39, Paris, March-April, 1980.

TÖDTLI, J., "Contribution to the Problem of Urban Road Traffic Control During Rush Hours", 2nd IFAC/IFIP Symposium on Traffic Control and Transportation Systems, Monte-Carlo, Sept.1974.

TOUGH, J.B., "MAXCAP A Method of Producing Optimal Signal Settings for Urban Networks During Peak Periods", Journal Inst. Math. and its Appl., Volume 15, Nov/Dec. 1979.

VAUGHAN, R.J., "Some Factors Affecting the Capacity of a Series of Signalised Intersections", T.E. and C., October 1975.

VINCENT, R.A., "Area Traffic Control in Bern", TEC, February 1972.

VINCENT, R.A. and LAYFIELD, R.E., "Nottingham Zones and Collar Study", TRRL Report 805

VINCENT, R.A., et al, "Bus Actuated Signal Control at Isolated Intersections, Simulation Studies of Bus Priority", TRRL Lab Report, 814.

WAGNER, F.A., "TRANSYT Method for Network Optimization", presented before the Institute of Traffic Engineers Annual Meeting, New York, New York, September 1972.

WEBSTER, F.V., "Traffic Signal Settings", Road Research Technical Paper, No. 39, Road Research Laboratory, England 1958.

WEINBERG, M.I., and GOLDSTEIN, H., "Digital Computer Controlled Traffic Signal System for a Small City", NCHRP Report 29, 1966.

WILLUMSEN, L.G., "Estimation of an O-D Matrix from Traffic Counts - A Review", Working Paper 99, Inst. for Transport Studies, University of Leeds.

YAGODA, H.N., "The Control of Arterial Street Traffic", presented at the IFAC Control Conference, Haïfa, Israël, 1967.

ZOVE, P., et al, "Integrated Motorist Information System (IMIS) Feasibility and Design Study, Phase I : Feasibility Study. Vol. 1. Final Report, Vol. 2. Appendices, Vol. 3. Executive Summary",U.S. Department of Commerce, National Technical Information Service, 5285, Port Royal Road, Springfield, Virginia 22161, PB Nos. 276090, 276091, 276092, April 1977.

ZUYLEN, H.J. van, "Congestie, Kosten Modellen, Bestrÿding", National Traffic Academy, Tilburg, Holland.

APPENDIX A

TERMINOLOGY

Bus lane - A street or highway lane intended primarily for buses, either all day or during specified period, may be used by other traffic under certain circumstances, e.g. right turns.

exclusive - A lane reserved for buses on a street, highway, bridge, or tunnel, which other traffic is restricted from using.

preferential - See bus priority system.

Reserved - See bus lane, exclusive.

Reversible - A lane that is reserved for the exclusive use of buses and other high-occupancy vehicles and that can be operated in either direction during peak hours.

Bus priority system - A means by which buses are given an advantage over other traffic, e.g., preemption of traffic signals or bus priority lanes.

Bus way - A special roadway designed for exclusive or predominant use by buses in order to improve bus movement and bus passenger travel times ; it may be constructed at, above, or below grade and may be located in separate rights-of-way or within highway corridors.

Commuter - A person who travels regularly between home and a fixed work or school location.

Contraflow lane - A highway or street lane on which, during certain hours of the day, public transport or other specially designated vehicles operate in a direction opposite to that of the normal flow of traffic on that lane during the remainder of the day.

Control of access -

Full control of access means that the authority to control access is exercised to give preference to through traffic.

Partial control of access means that the authority to control access is exercised to give preference to through traffic to a degree that there may be some crossing at grade and some private driveway connections.

Corridor - System of transport facilities built to provide high level of transport in a linear stretch of land of several kilometres.

Cycle -

Cycle-length - The total time required for a complete sequence of signal phases.

Delay - The time consumed while traffic or a specified component of traffic is impeded in its movement by some element over which it has no control. Usually expressed in seconds per vehicle.

Fixed delay - The delay to which vehicles are subjected during light traffic volumes or low densities. The delays experienced by a single vehicle as a result of traffic signals or stop signs are fixed delays.

Operational delay - The delay difference between travel times over a route during extremely low and high traffic volumes is operational delay. The time consumed while waiting at a stop sign for cross traffic to clear, the time losses resulting from saturation, from interference with parking vehicles, and from turning vehicles are examples of operational delays.

Density - The number of vehicles occupying a unit length of the through traffic lanes of a roadway at a given instant. Usually expressed in vehicles per kilometre per lane.

Average density - The average number of vehicles per unit length of roadway over a specified period of time.

Critical density - The density of traffic when the volume is at capacity on a given roadway. At a density either greater or less than the critical density, the volume of traffic will be decreased. At critical density all vehicles are moving at more or less the same speed.

Diversion - The shifting of traffic from one mode of travel to another or from one route to another.

Flow - Volume - The rate at which vehicles pass a point of a road per unit of time.

Follower - All vehicles except the first one, the leader, in a moving queue (platoon or bunch).

Gap - The time or space interval between individual vehicles measured from the rear of one vehicle to the front of the following one.

Headway - See gap.

Impedance - The conditions, such as capacity limits, high costs and distance, that may restrict travel over a transport link.

Lane, traffic - A strip of roadway intended to accommodate a single line of moving vehicles.

Reversible lane or lanes - A lane or lanes where traffic moves in one direction only during some period of time, then in the reverse direction during another period of time.

Leader - Vehicle situated at the head of a platoon, or bunch.

Mixed traffic - Traffic composed of different modes sharing the same lane (e.g. motor vehicles - bicycles, buses - motor vehicles).

Modal split - 1. The proportion of total person-trips per specified modes of transport. 2. The process of separating total person-trips into modes of travel used.

Mode - A means of travel, e.g. automobile, public transport or walking.

Door-to-door - A single mode of transport that provides service between the exact origin and destination points without transfer or use of another mode.

Network - 1. A system of links and nodes that describes a transport system. 2. The configuration of highways or public transport routes and stops that constitutes the total system.

Node - A point that represents an intersection of two or more highways or public transport routes or a zone centroid used in traffic assignment.

Off-line - Not in the main flow of traffic or processes.

Offset - The number of seconds or percent of the cycle length that green appears at a given traffic control signal after a certain instant used as a time reference base.

On-line - 1. Under direct computer control. 2. In the main flow of traffic or processes.

Phase - A part of the time cycle allocated to any traffic movement or to any combination of traffic movements receiving the right-of-way simultaneously during one or more intervals.

Platoon - A closely grouped elemental component of traffic, composed of several vehicles moving with clear spaces ahead and behind.

Priority lane - A lane reserved for a selected group of vehicles, i.e. buses, car, or van pools.

Progression, signal - A signal system in which the successive signal phases controlling a given street give "go" indications in accordance with a time schedule to permit (as nearly as possible) continuous operation of groups of vehicles along the street at a planned rate of speed, which may vary in different parts of the system, or according to time of day.

Quality of service - 1. Multidimensional characteristics that indicate the quality and quantity of transport service provided, including characteristics that are quantifiable (travel time, travel cost, number of transfers) and those that are difficult to quantify. 2. A measure of the effectiveness of a highway in serving traffic in terms of operating conditions.

Queue - 1. A waiting line of vehicles, e.g., traffic at a signal, buses at a park-and-ride facility. 2. A waiting line of persons.

Moving - A queue of vehicles which move along a road, each one following the movement and reacting to the actions of the one ahead ; see platoon, follower, leader.

Rerouting - See diversion.

Sensor -

Discrete - A sensor that does not provide qualitative information, but has a binary output, i.e. "on" or "off".

Induction loop - A loop of wire (inductor) embedded in the roadbed that carries a small electric current that is used to sense a passing vehicle and to yield information about the presence of the vehicle.

Split - Green split is the distribution of cycle time over the phases.

Throughput -

Passenger - The number of people who pass a given point or set of points during a given period of time.

Vehicle - The number of vehicles that pass a given point or set of points during a given period of time.

Traffic control system, centralised - A traffic control system in which the signals and switches for a designated section of track are controlled from a remotely located centralised traffic control machine.

Trip assignment - A process by which trips described by mode, origin, destination, and time of day are allocated among the paths or routes in a network according to a flow distribution model.

Capacity restraint - An iterative process by which the volume allocated to a route is compared with the capacity of that route and the speed of the route is adjusted accordingly to reflect its characteristics of speed, volume, and density ; new minimum-time paths are calculated at the beginning of each iteration.

APPENDIX B

VEHICLE - AND TRAFFIC - ACTUATED CONTROL MODELS AT ISOLATED INTERSECTIONS

This Appendix provides a brief summary of existing methods of providing traffic responsive control at individual isolated intersections. In general, the term "Vehicle-Actuated Control" infers some form of control strategy which is directly dependent on measured vehicle activity.

Traffic actuated control is used when data of all traffic is available to calculate the signal phase on-line.

B.1. CONTROL PARAMETERS

One or more of the following traffic flow characteristics usually forms the basis of traffic responsive signal control.

1. Vehicle interval : Traffic volumes on the controlled approaches are measured by small area detectors and green time is assigned to each movement sufficient to serve the measured demand (a pre-selected amount of green time for each vehicle). One or more phases may be responsive to traffic flow.

2. Vehicle waiting time (gap reduction) : This method takes into account the length of time that vehicles are waiting to be served. This is accomplished by reducing the amount of green time given to each unit being served as clock time passes. (The vehicle interval described in item 1 above is reduced over a sliding scale.)

3. Volume density : This method, a variation of item 2 above,relates traffic volume and density of flow (time between successive vehicles) in determining the amount of green time for each phase.

4. Lane occupancy : This method involves the use of a "long loop" placed on the approaches (sometimes in individual lanes). Green is extended for any phase while a vehicle is in the "long loop". It is essentially a control mode where a distance is used instead of a time period.

5. Rate of occupancy : This method takes into account the volume of vehicles waiting to be served.

B.2. CONSTRAINTS ON TRAFFIC-RESPONSIVE CONTROL

When utilising any of the above parameters as the basis for control several conditions must be considered, each of which may limit the flexibility of the strategy.

1. Maximum green times : The length of green time may be limited due to practical considerations. Controllers are available, that provide different settings for different conditions.

2. Pedestrian crossing time.

3. Safe stopping distance (high speed approaches).

4. Queue length and storage area.

5. Intersection geometry, i.e. clearance times.

B.3. EVALUATION

A number of both field experiments and computer simulation projects have been undertaken in an attempt to evaluate the above strategies of responsive type control :

1. Various studies in the United States, France and Australia, have not shown any agreement as to the superiority or other wise of any one of the above strategies.

2. Miller has proposed a strategy based on minimum delay which differs from the above empirical techniques in that it is "adaptive" rather than pre-set. This is the result of a traffic-actuated overall approach instead of a vehicle-actuated adaptive approach.

3. Simulation by NIP showed little difference between Miller's technique and conventional vehicle-actuated control at moderate volumes. However Miller's resulted in less delay than did fixed time control at volumes near saturation.

4. A Miller-type control strategy known as T.O.L. has been tested in Sweden using various detection configurations with satisfactory results (Traffic Optimisation Logic). These tests also showed the need for accuracy in the control.

5. A French simulation and field study undertaken by IRT has shown that the Miller method is more efficient than a vehicle-interval strategy which in turn is superior to fixed cycle control. However the latter was more efficient than the vehicle-interval control mode in the case of saturated traffic conditions.

6. Dynamic programming was used by Robertson to determine the optimum control of an isolated traffic signal with predetermined but arbitrary flows. The DYPIC method could be used to establish an absolute standard of performance against which the merits of practical control policies can be judged.

Some conclusions may be stated based on these studies :

1. Basic Vehicle Interval control is more efficient than fixed time control for unsaturated conditions. It is not obviously more efficient at saturation.

2. Many strategies require advance information. The configuration and location of detectors is therefore vital.

APPENDIX C

ON-LINE MODELS

C.1. UTCS (United States)

The three generations of the Urban Traffic Control System have all been designed for centralised computer control with a situation-dependent amount of freedom for variations on the basis of intersection data.

First Generation Control System

The first generation software uses prestored timing plans developed off-line and based on previously generated traffic data. The system can store up to 40 timing plans per section. A section is a group of local controllers operating in the same mode and switching simultaneously from one timing plan to another. Timing plan selection of a desired set of timing parameters (cycle, split, offset) may be based on time of day, automatic response to traffic conditions, or operator choice. The mode of timing plan selection is determined by the operator. Four modes of area control are available in 1-GC :

1. Standby (STBY) - Controllers are transferred from computer to local backup system control.

2. Time of Day (TOD) - Timing plans are automatically selected on a time-of-day and day-of-week basis. Change to the selected plan is accommodated with a time resolution of 15 minutes for each section in the system. This mode may be selected by the operator on a system basis (all sections), or on an individual section basis.

3. Traffic Responsive (TRSP) - This mode automatically matches the timing plan best suited for existing traffic conditions. Very recent volume and occupancy data are compared with volume and occupancy characteristics of the various stored patterns. Each history for a section of controllers is unique to a set of control parameters (controller timing plans). The control plan related to the best data-history match is selected from the available timing plans.

4. Manual (MAN) - This mode is used primarily for handling unusual traffic conditions not provided for when operating under the TOD and TRSP modes. The MAN mode may be used during checkout of traffic-responsive control patterns or to provide rapid response to unusual traffic conditions resulting from parades, concerts, or sporting events.

For the oversaturation aspect the Critical Intersection Control is introduced. Critical intersections which saturate frequently may have their split adjusted to be directly proportional to the traffic demand for green time computed for their associated links. The cycle splits for these intersections are then changed on a once-per-cycle basis as a function of the values of queues and volume on these links. This critical intersection control (CIC) is used to fine tune the split allocated to each approach to the intersection, based upon fluctuations in local traffic demand. Thus, intersections control works in conjunction with, and supplements, area control operation.

Second Generation Control System

The second generation control (2-GC) strategy is a realtime, on-line system that will compute and implement signal plans based on surveillance data and predicted changes. The 2-GC program, designated TANSTP, (Traffic Adaptive Network Signal Timing Program), retained the selection of prestored timing plans through the manual or time-of-day modes, and the hardware failure detection functions of the 1-GC UTCS software and hardware. The optimisation process is repeated at 5 minute intervals. New timing plans could not be implemented more often than every 10 minutes to avoid developing control parameters based on conditions which exist during signal timing transition.

The 2-GC software provides the following features :

1. Optimisation algorithm - The OPTIMIZ module of the SIGOP program was selected because of its short execution time, moderate computer memory storage requirement, and its simple and explicit mathematical optimisation function (a quadratic equation of stops and delay terms).

2. Traffic prediction model - Selected to predict volume and speed as a function of historical volume and speed, so that computed timing plans do not slow down current flow conditions.

3. Subnetwork configuration models - The 2-GC software can improve optimisation efficiency by dynamically reducing the total network into subnetworks on the basis of prevailing traffic conditions.

Again for oversaturation the critical intersection control is used. The CIC routine for 2-GC involves changing the cycle splits and offsets for certain intersections once per cycle as a function of the demand (computed values of queues and volume on these links).

Third Generation Control System

The third generation control (3-GC) UTCS software was developed to implement and evaluate a fully responsive, on-line traffic control system. The 3-GC software was designed to permit the cycle length of each controller to vary from cycle to cycle. This design approach differentiates 3-GC from those control systems that provide a sequence of fixed-time, cycle-based signal timings. The 3-GC software system interfaces with existing field controller hardware; the local dial system therefore remains intact and provides a back-up (standby mode) for the computer-based system. Both 1-GC time-of-day and manual modes can be used.

The operational features of 3-GC software provide for the following :

1. Signal timing is computed for all controllers at least once every 3 1/2 minutes.

2. Signal offset and split are determined to minimise vehicle delay and stops along each approach and provide networkwide coordination.

3. Congestion paths are identified and the control algorithm is specifically designed to service a congested condition while the condition persists.

4. Transition routines are not needed because the system is constantly in a state of transition as it responds to changing traffic conditions.

In the 3-GC traffic-responsive mode, signal timing plans are developed by either of two optimisation algorithms : The Cycle-free Responsive Algorithm for Network Optimisation (CYRANO) provides signal timing in undersaturated conditions, and the Congested Intersection Control coupled with Queue Management Control (CIC/QMC) provides areawide signal timing for saturated conditions on congestion paths.

Signal co-ordination is accomplished in CYRANO by implementing a coarse simulation of traffic flow and then systematically adjusting the signal settings at each controller to minimise disutility (a linear combination of vehicle stops and delay aggregated over all approaches). The objectives for the CIC/QMC are to maximise throughput at the congested critical intersections to provide acceptable service for the two congested competing demands and to manage queue formation along the paths leading to the critical intersection so that the intersection is not blocked and green time is fully utilised.

C.2. PBIL On-line tests in Aachen (Germany)

Basically PBIL is a control strategy for a "star" of one critical intersection plus the related intersections around it.

The algorithm is based on traffic data in all approaches of the controlled critical intersections. Constraints are the minimum green and the maximum red settings but also maximum queue length. Further input data are split and offset of adjoining intersection.

PBIL tries to minimise losstimes which are defined as overtraveltime in the network compared with the theoretical normal traveltime. Losstimes contain delay and stops. When certain critical queue lengths are reached, losstimes are increased artificially in order to prevent oversaturation.

The general control strategy for PBIL can be described as follows :

- identification of critical intersections,

- separation of these locations from the system when traffic flow is nearly saturated or unbalanced,

- operation of these areas with PBIL.

The number of PBIL controlled intersections in a network depends on hardware aspects.

Under non-saturated conditions the effect of a change of phases is calculated in the way originally proposed by A.J. Miller. The saturation of a link will lead to a queue. When the length of this queue exceeds a certain critical value the parameters in the calculations are artificially changed to favour the congested link.

C.3. SCOOT (United Kingdom)

It operates groups of adjacent junctions on a common cycle time. At any instant, the cycle time, green durations and offsets between signals are controlled by timings held in computer store. SCOOT estimates how to adjust these timings to reduce total delay and stops. Frequent small alterations adapt the signals to short-term fluctuations in the traffic demand. Longer-term trends are satisfied by the accumulation of small changes over several minutes. Thus, there are no large sudden alterations in timings that might disrupt traffic flow. Special procedures deal with congestion ; on every link queues can be measured. Once congestion is detected the normal optimisation programs will favor the congested link thanks to a special term used in the optimisation. However, no extra green will be given when the extra traffic would block the junction. Initially, a traffic simulation program was used to compare alternative strategies. The best strategies were implemented on the computer in the Urban Traffic Control Centre in Glasgow, Scotland.

The Glasgow UTC System controls 96 signalled junctions. To evaluate SCOOT, inductive loop vehicle detectors were installed on the approaches to 40 junctions in the central area. The detectors were connected over data transmission lines to the computer. A suite of real-time computer programs was developed to implement the SCOOT strategy of control. Trials to measure the effectiveness of SCOOT took place during 1975.

The SCOOT traffic-responsive method of control was compared with fixed time plans derived by TRANSYT. In previous experiments, TRANSYT improved average journey speeds by about 16 per cent compared to the conventional methods of control used prior to installation of a computer.

New TRANSYT plans were calculated for the trial. SCOOT and TRANSYT were compared by measuring average journey speed using four specially instrumented cars. Measurements were made over 15 working days for both methods of control.

Results in Glasgow show that SCOOT improved journey times by 6 per cent throughout the working day. Experiments in Coventry have borne out the results obtained in Glasgow.

As far as saturation is concerned SCOOT thus works in a preventative way. As soon as traffic forms queues extra green can be given. But as soon as there is a danger of spillback flows will be restricted, thus redistributing demand to an upstream mode.

C.4. SCAT (Australia)

In New South Wales, Australia, a new traffic control system has been designed and implemented using microprocessor local controllers and minicomputers for networks (up to 200 intersections). On top one central computer is used for monitoring and for data collection. The control program developed is called SCAT Sydney Coordinated Area Traffic system.

SCAT is a traffic-responsive system. Vehicle detector information is used for both local VA (tactical) control and for plan and cycle-length selection on an area basis (strategic). Cycle length is continuously variable, cycle by cycle, whilst splits and offsets vary in steps (by plan change) and continuously (by modification of the plan on the basis of cycle length and system activity level). Defined sub-systems may operate asynchronously or, when traffic conditions demand, linked together on a common cycle length with a defined offset between sub-systems.

In light traffic conditions it is usual for SCAT to minimise stops, in medium traffic to tend more to minimise overall delay, whilst in congested conditions, splits and offset tend towards maximising throughput on major routes.

Part of the program is the control of congestion. SCAT, through its detectors, recognises the onset of congestion and drastically alters splits in an effort to minimise interaction between controlled intersection (i.e. to put queues where they are least likely to cause system breakdown). At the same time offsets are altered to suit maximum throughput.

Cycle-length tends to maximum to maximise capacity. Input to the system may also be restrained.

C.5. GERTRUDE (France)

The program GERTRUDE specially designed for areas with congestion, is used in Bordeaux, France.

In principle, the system works with a fixed emergency procedure that is used to prevent congestion. The measures are therefore made effective as soon as congestion is detected.

Around every critical intersection there is a small area with a number of intersections. The Bordeaux system relies on the control of the main roads per area of zone. To prevent a blocking of intersections (spillback) the program will not allow traffic to block a junction. It will rearrange the green split in such a way that the critical intersection has a maximum throughput at the expense of one link, the storage link that has to buffer all the excess demand. These storage links are now on the main approach roads.

C.6. RTOP (Toronto, Canada)

A full-scale digital traffic control system has been in operation in Metropolitan Toronto, where, since 1963, the concept is pioneered. In-house research activities during the interveening years have resulted in extensive modifications and the system is presently controlling approximately 1200 signalized intersections.

A new Real Time Optimisation Package (R.T.O.P.) is currently being brought on-line to supersede the original system. This program will include three levels of control and is designed to accommodate 2000 signals (at any one time it is estimated that some 40 % of the intersections will be under the R.T.O.P. mode of operation). The following is a brief description of the Program.

Configuration

1. Detectors will be provided at the rate of approximately 4 per intersection (one per link).

2. Thirty-two individual intersections each controlled by a micro-processor type controller are linked to form a cell.

3. Each cell will be controlled by a 32 K mini Computer.

4. The cells in turn are controlled on a tactical basis to :
   - determine cycle length
   - call for traffic-responsive operation
   - override local control and control the cell.

5. Strategy control is also provided determining new plans to be implemented every 15 minutes.

Input Requirements

Two types of data are utilised, i.e. real-time data from the loop detectors and off-line data from the following :

1. Vehicular turning movements (percentages by 15-minute periods).

2. Historical flow data for all links for the last several days.

3. Data to describe the network (speed, detectorlocations, etc.).

Strategy

The Real-Time Optimisation Program (RTOP) package is a modular set of computer programs designed to compute optimal signal timing parameters (cycle lengths, offsets and splits) on a real-time basis, in response to the changing on-street traffic volume patterns. It is a second generation control technique and as such computes the signal timings for a finite control period as compared with third generation control techniques which respond dynamically on a cycle-by-cycle basis.

1. Prediction

   As is the case for any real-time optimisation procedure, traffic volume prediction is a key component.

   The traffic flow model in the RTOP package is based on the prediction logic developed for use in the ASCOT (Adaptive Signal Control Optimisation Technique) program, with the addition of certain modifications and refinements. The model predicts, for each fifteen-minute control period, traffic volumes for each detectorised link taking into account three prediction components, namely, historical volume trends (day-to-day), current volume trends (control period to control period) and directional relationships (volume trends on a control or representative link). These three estimates are combined to form an overall volume prediction which is then tested for consistency with past data. Since it takes a finite time to complete the prediction process and the subsequent signal timing computation, a time lapse is inevitable between real-time data acquisition and signal timing implementation. However, the use of the three prediction components minimises the effect of this time lapse as much as possible. The same prediction procedure is applied to both detectorised links with real-time traffic counts and non-detectorised links with counts extrapolated from adjacent detectorised links.

2. Signal timing for individual intersections

Once the traffic prediction process has been completed for all the links in the network the RTOP package computes the appropriate signal timings in response to the predicted traffic demand.

First, each individual signalised intersection is considered and a number of possible cycle lengths are calculated on the basis of the following design criteria :

(a) Absolute minimum cycle length : taking into account minimum pedestrian walk time and clearance interval requirements for each phase.

(b) Cycle length based on Webster's Method.

(c) Cycle length based on minimum green and vehicular flows for each phase.

Based upon the above assumptions, two final cycle lengths are identified for each individual signalised intersection ; one a practical minimum and the other a practical optimum.

3. Cycle-length selection (network)

Following the signal timing computations, a range of candidate system cycle lengths are systematically tested and evaluated to find the "best" cycle length for the system.

For each the RTOP package performs the following general functions :

(a) Selection of the intersections which will be under co-ordinated control in the system and those which will be isolated from the system : The criterion for this selection is that when the candidate system cycle length is less than the individual minimum intersection cycle length, this intersection is placed under isolated control.

(b) Calculation of the signal-timing intervals for both the co-ordinated and the isolated signals in the network.

(c) Evaluation of the performance of the non co-ordinated signals in the network in terms of delay and stops, using the technique developed by Webster.

(d) Computation of delay/difference-of-offset relationships using the signal and volume data from the upstream signal, then link exist flow pattern at the downstream signal.

(e) Identification of the ideal offset setting for each link based on a performance index computed for each given offset difference between the upstream and downstream signal. This performance index is based on the amount of delay, resulting from any particular offset difference.

(f) Calculation of link importance factor, being analogous to the sensitivity of its performance towards the choice of offset difference. For a highly important link an offset setting close to the ideal value should be implemented, whereas for an unimportant link the use of the ideal offset setting is not essential.

(g) The optimum offset (the offset values associated with the global minimum system performance index) is calculated.

(h) System Performance Evaluation is obtained by adding the performance index from the isolated signals and the performance index for the co-ordinated signals corresponding to the calculated offsets. The cycle length, signal phasing, and offset settings corresponding to the lowest overall system performance index are finally chosen for implementation.

4. Implementation

For each fifteen-minute control period, the program computes signal splits and optimises the system cycle length and the associated offset settings on a real-time basis in response to changing traffic conditions in a computer-controlled signal network. In addition, the co-ordination structure of the network is re-configured in every control period based on various cycle length alternatives.

APPENDIX D

OFF-LINE MODELS (STUDY TOOLS)

The off-line network control models (computer programs) can be summarised as follows :

D.1. TRANSYT

The most common model is the Traffic Network Study Tool TRANSYT developed by T.R.R.L. This model consists of four parts. The first part checks the input data given by the programs. This input data describes the streets in terms of links that carry traffic in one direction and nodes or intersections. The input also describes how much traffic is following a link and from which other links at the node downstream traffic comes. The restraints like minimum green and clearance time are given per traffic-light controlled node. Also information on the length of the link and the free flow travel speed is given in the input. One can indicate whether two links merge or use a shared stop-line at their node at the upstream. Finally, one can indicate in which way and how accurately TRANSYT should try to find the optimum traffic light settings for each node. A very strong point in favour of TRANSYT is the accuracy with which the input data are scrutinized.

The second part of the program is the simulation part. It simulates the traffic on the links during one full cycle. This is done by splitting up the cycle in time steps of equal length. Then the traffic situation, i.e. the number of vehicles in front of each stopline (moving or waiting) is calculated. This program also calculates the value of the parameters used for the representation of the performance. These can be delay, bus delay, stops, or in special TRANSYT versions, emission etc. This information combined as a weighted sum in the "Performance Index" is indicative for the given situation, thus for the light setting used.

The third part of the program is the optimisation part. It uses a hill climb technique that step-by-step changes the setting of a traffic light and that optimises on the effect this change has on the Performance Index.

The fourth part is the output part. This will, if required, produce an enormous amount of data including graphs showing how traffic behaved at a stopline. This information enables the traffic engineer to use the program as a Study Tool. It makes it possible to understand what happens and thus it allows him to change the traffic light setting in such a way that he achieves his goal.

By its very nature this TRANSYT program is not able to optimise oversaturated conditions. There are three problems that have to be solved before one can use the standard TRANSYT. First details in the delay calculation model have to be changed. Secondly, details in the traffic behaviour on the links have to be changed. Thirdly, the distribution of demand over time has to be taken into account. However, a very carefull application of TRANSYT to study preventative measures is possible.*

D.2. SIGOP II

This is a program developed for the Federal Highway Administration for the purpose of computing optimum signal timing plans for either arterial or network systems. The program is composed of two major components :

1. A flow model to describe traffic behaviour compatible with the applied signal timing,
2. An optimisation scheme designed to yield signal timings which service current patterns of traffic with maximum efficiency.

Traffic is modelled as a platoon of vehicles which disperses as it travels along a link. The effect of traffic control at an intersection is measured by a change in disutility, which is derived from changes in volume and platoon structure of traffic departing from an intersection. For multiphase control, the platoons are stratified by turning movements.

---

* Transyt 8, the latest version, however, has been adapted so as to handle oversaturated conditions.

Signal timings (splits and offsets) are obtained from a process whereby a series of optimisation problems are solved for every node in a network. Mini-networks are constructed around each node ; control settings at peripheral nodes are frozen and estimates of delay, stops, and queue length are obtained from algebraic relations expressed in terms of control settings and platoon structure. These estimates are combined to represent the disutility measure, and the signal timings that minimise this measure for each mini-network are identified through the use of a gradient search technique. All nodes in the network are examined in this manner. Further iterations examine changes in the on/off times of phases at each node to account for changes which have been made at the surrounding peripheral nodes during the previous iteration. Iterations proceed until a minimum network-wide disutility value is reached.

Of special interest to this report is the addition made to handle oversaturation :

An intersection is considered to be saturated in SIGOP-II when the calculated cycle length (sum of the required green times for all phases) is 20 per cent greater than the input cycle length. When this condition occurs, the program will first attempt to adjust splits to give more green time to the saturated phase.

The program checks to see that the green times calculated by the program for each undersaturated phase are at least 10 percent greater than the minimum required green times. If this condition is met, one second is taken from each undersaturated phase and added the oversaturated phase. If this condition is not met or the situation has not been remedied, a modified Webster's equation will be used for estimating phase duration, provided that one of the undersaturated phases has a calculated green time which is 25 per cent greater than the required minimum phase duration. If this condition is met, the green time for that phase is reduced by an amount proportional to the difference in volume-to-service-capacity ratio between that phase and the oversaturated phase. This green time is then added to the oversaturated phase.

If all of the above procedures fail, or conditions are not met, then an adjustment is made to the calculation of offset to avoid over-saturation. The adjustment is actually made in the estimations of stops and delays which are used in the model's objective function to compute disutility. Offsets are adjusted to minimise the disutility measure.

Summarising : In the case of saturation first the split (of green) is adjusted; if that doesn't solve the problem coordinations are recalculated to avoid a spillback effect (filtering). As the situation is simulated during one cycle, when one wants to study oversaturated conditions in their relation to time varying demands, additional manual work has to be done.

D.3. SIGSET

SIGSET is a maximum bandwidth program. The problem with this type of program is that a large bandwidth is probably of little value when congestion is present. The reason is that the presence of queues, which inevitably form in heavy traffic, is not taken into consideration in designing for maximum bandwidth.

D.4. THESEE SATURATION

THESEE saturation is the special version for saturated conditions of the program. THESEE, like TRANSYT, represents the network in nodes and links. Like TRANSYT it has an optimisation program to optimise the traffic light setting on the basis of the effect the program measures when simulating traffic. The essential difference is that THESEE saturation is based on the philosophy or strategy of storage links. Storage links are preassigned links that will be used as buffer any time traffic cannot leave this link without creating saturation on the downstream network. Demand is thereby redistributed both in space and in time.

The choice of the storage links that can be entry links or internal links of the network can be made by the programmer in an iterative process. With a given set of storage links that each have a given green phase per storage link THESEE Saturation will adjust the other green phases so that a maximum performance expressed in terms of network flow or throughput is reached.

When using this model one has to study the effect of the application of different green phases.

Again the effect of time-varying demand cannot be taken care of by the study model itself. The model is still under study. It can be seen as a generalisation of the on-line system GERTRUDE used at Bordeaux.

APPENDIX E

MATHEMATICAL MODELS

E.1. INTRODUCTION

This Appendix will deal in greater detail with some theoretical approaches to the problems of saturation. As theoretical, one may classify any method where the control strategy is derived mathematically for a given optimisation criterion and under stated assumptions.

The following concentrates on a few, more typical, approaches while only briefly mentioning or omitting some others. This selection does not necessarily imply an evaluation as to the relative merits of the methods.

The approaches described have been classified in two main categories : pre-timed (or off-line) strategies and traffic-controlled (on-line) strategies. These categories are not mutually exclusive but rather the two ends of a spectrum. Some traffic-controlled modes make use of pre-programmed flow predictions and pre-timed plans but could also be based on actual traffic conditions as input data. The choice of method is in some way related to the choice of criterion. A queue length criterion applies to all situations at any one moment in time, while average delay obviously has to refer to an extended time interval with normally non-stationary input flows.

E.2. PRE-TIMED STRATEGIES

During normal, non-saturated conditions, the best fixed time control is calculated using arrival flows Q_i (i = 1,2.. number of lanes) which typically represent average values for a period of one hour or more. As long as the arrival flows do not reach capacity for any substantial part of that period, this averaging will have little effect on the performance of the intersection in terms of mean delay, for example.

If capacity is exceeded, however, it might take some considerable time before the accumulated queues are again discharged and normal conditions restored. See Figure E.1, which shows that the queues are at their longest at the moment the arrival flow goes back to sub-capacity level. The figure represents one approach and the saw tooth curve caused by the signal has been smoothed out. (Random delays are ignored).

Figure E.1

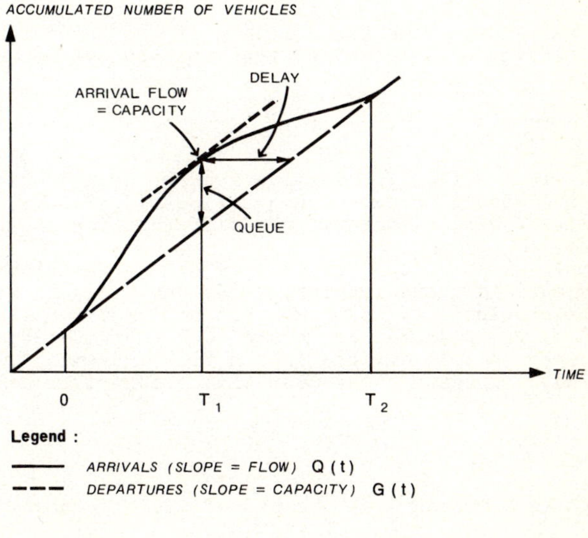

In the example of Figure E.1, it also appears that the best control during some short time interval (t, t + dt) should consider not only the arrival flows $Q_i(t)$ at that time but also the future developments of these flows, assuming they are known.

Gazis and Potts choose as optimising criterion aggregate delay D during the time interval (O,T) and formulate the problem as follows : They consider an intersection with two one-way streets and no turning traffic, i.e. i = 1,2.

$$\min D = \sum_{i=1}^{2} \int_{0}^{T} [Q_i(t) - G_i(t)] \, dt$$

$$Q_i(t) = \int_0^t q_i(\tau) d\tau \qquad \text{(cumulative demand)}$$

$$G_i(t) = \int_0^t \gamma_i(\tau) d\tau \qquad \text{(cumulative service)}$$

Random variations are neglected and the saw tooth curve is smoothed out so that the following boundary conditions apply (as in Figure E.1) :

$$Q_i(0) = G_i(0)$$

$$Q_i(T) = G_i(T)$$

$$Q_i(t), \quad G_i(t) > 0, \quad 0 < t < T$$

Cycle time is assumed constant (c) and the service function will, thus, be related to green times $g_i(t)$ through $\gamma_i(t) = s_i \, g_i(t)/c$

s_i = saturation flow (constant).

The green times g_i are the control variables restricted by :

$$g_1(t) + g_2(t) = c - L \qquad \text{(L = lost time per cycle)}$$

$$g \min \leq g_i \leq g \max$$

where minimum time g min and maximum time g max are determined by practical considerations (such as pedestrian safety and public acceptance).

The problem stated may be solved analytically, for example by using Pontryagin's control theory as shown by Gazis. Gazis discusses a number of examples and also extends the problem to two interacting intersections (neglecting travel time). In general, the best policy for a single intersection consists of a two-stage operation ("bang-bang").

(1) let g_1 = min (gmax c - L - gmin), $0 \leq t < T'$

(2) let g_2 = min (gmax c - L - gmin), $T' \leq t < T$

(It is assumed that $s_1 > s_2$).

If the objective was to minimise the duration of the saturation period (T), the "switchover" point T' should be selected so that both queues are emptied simultaneously (at T). This happens if

$$T' = \left[\frac{c}{s_1} Q(T) - g\min T\right] / \left[g\max - g\min\right]$$

where T is such that Q (T) = G (T). To minimise average delay, however, the maximum service rate for direction 1 should be extended a bit beyond this switchover point. In that case, the time (T) when saturation ends will be different for the two approaches.

To facilitate numerical solutions for general demand functions, and also to admit considerations such as maximum queue lengths, d'Ans and Gazis as well as Church and Revelle state the above problem in a form amenable to linear programming techniques. They consider a time period T, which should be chosen such that it certainly includes the time of saturation. If this period T is divided into n time increments, each of a duration of one cycle (i.e. n = T/c), aggregate delay over the period (O,T) can be approximated as :

$$D = \frac{c}{2} \sum_{i=1}^{2} \sum_{j=1}^{n} x_{ij}$$

x_{ij} = queue length in approach i at the end of cycle j (state variables)

We also have to smooth out the arrival function by defining

$$a_{ij} = \int_{(i-1)c}^{ic} q_i (t)\, dt = \text{number of arrivals in approach}$$

i during cycle j (where q(t) is the arrival rate)

The effective green times g_{ij} for cycle j = 1,2.. n are control variables and can be regarded as having two components :

$g_{ij} = w_{ij} + u_{ij}$ where

w_{ij} = part which is used by queueing vehicles

u_{ij} = slack part of effective green time

Since cycle length is taken to be fixed, the following must hold :

$$\sum_{i=1}^{2} g_{ij} = \sum_{i=1}^{2} w_{ij} + \sum_{i=1}^{2} u_{ij} = c - L$$

L = lost time per cycle

The number of cars discharged during cycle j will be $w_{ij}.s_i$, where s_i = saturation flow for approach i. The queue length at the end of the cycle can now be expressed as the difference between total number of arrivals and total number of departures for cycles 1, 2 .. j :

$$x_{ij} = \sum_{k=1}^{j} a_{ik} - \sum_{k=1}^{j} w_{ik}\, s_i$$

To summarize, the complete LP - formulation will be as follows :

$$\text{Minimise } D = \frac{c}{2} \sum_{i=1}^{2} \sum_{j=1}^{n} x_{ij} \qquad (1)$$

subject to

$$\sum_{i=1}^{2} w_{ij} + \sum_{i=1}^{2} u_{ij} = c - L \quad (j = 1,2..n) \qquad (2)$$

$$x_{ij} + \sum_{k=1}^{j} w_{ik}\, s_i = \sum_{k=1}^{j} a_{ik} \quad (j = 1,2..n\ ;\ i=1,2) \qquad (3)$$

$$x_{ij},\ w_{ij},\ u_{ij} \geq 0 \quad (j = 1,2..n;\ \ i=1,2) \qquad (4)$$

The problem is easily solved using the Simplex method or some of the more modern techniques such as the sub-gradient method.

Church and Revelle suggest the introduction of some further constraints. Due to the "bang-bang" effect described above, restrictions on minimum and/or maximum green would be desirable or else one approach can remain red for a considerable number of cycles. Although average delay will increase, it might also be desirable to impose a restriction on maximum individual delay, such as :

$$\sum_{k=1}^{j+h} w_{ik} - s_i \geq \sum_{k=1}^{j} a_{ik} \quad (j = 1,2..n\ ;\ i = 1,2) \qquad (5)$$

in which case no vehicle will have to wait more than h cycles. Further, the maximum queue length can be restrained through

$$x_{ij} = X_i \quad (j = 1,2..n\ ;\ i = 1,2) \text{ where} \qquad (5)$$

X_i = maximum queue length in approach i.

Restrictions (5) and (5') must not be chosen too severely, or it could happen that the problem has no solution.

An alternative problem, and a quite different solution, is obtained if the objective is to minimize maximum queue length X instead of aggregate delay. Using the same notation, this problem can be stated as :

$$\text{Minimise } X \tag{6}$$

subject to

$$x_{ij} - s \leq 0 \qquad (j = 1,2..n \; ; \; i = 1,2) \tag{7}$$

$$x_{ij} + \sum_{k=1}^{j} w_{ik} \, s_i = \sum_{k=1}^{j} a_{ik} \qquad (j = 1,2..n \; ; \; i = 1,2) \tag{8}$$

$$\sum_{i=1}^{2} (w_{ij} + u_{ij}) = c - L \qquad (j = 1,2..n \; ; \; i = 1,2) \tag{9}$$

$$x_{ij}, \; w_{ij}, \; u_{ij}, \geq 0 \qquad (j = 1,2..n \; ; \; i = 1,2) \tag{10}$$

Constraints on minimum green, maximum green or maximum waiting time can be added. This problem is also linear and easily solved. The extension of the model to two interacting intersections without turning phases, as discussed by Gazis and d'Ans, Burhardt and Michalopoulos and Stephanopoulos, is relatively straightforward. As shown by the latteer two references, the effect of travel time can easily be accommodated. Michalopoulos and Stephanopoulos use continuous time which leads to a more complicated technique, although the results are much the same.

For a more general network, however, the problem will be linear only if all origin-destination pairs along a link can be assumed to be independent, that is be provided with separate lanes. If one link has to handle several OD-pairs (m = 1,2..) in one and the same queue, some assumption has to be made regarding the order of service. Gazis and d'Ans suggest that the discharge rates $_{n}{}_{ij}{}^{k}$ (t) for link ij towards node k for OD-pair m should be considered proportional to the respective queue sizes at each and every cycle. This assumption, however, causes non-linearity for all cases with more than one OD-pair.

Again, for a single intersection, it should be noted that the formulations above all assume a constant cycle time. This cycle time may be determined iteratively, but would cause non-linearity if introduced in the optimisation of splits. This non-linearity can be eliminated if arrival flows are assumed constant, as in the SATURNE method of David and Henry. For an isolated intersection, their criterion of optimisation is :

$$\frac{1}{n} \sum_{i=1}^{2} \sum_{j=1}^{n} x_{ij}$$

Cycle times c_j and green times g_{ij} (for phase i = 1,2) are obtained through linear programming. Since cycle time varies, the criterion gives only an approximate solution to minimum aggregate delay. A possibly more coarse approximation follows from the assumption $g_{1j} + g_{2j} = c_j$, i.e. lost time is disregarded.

To summarise, theoretical solutions to the stated off-line control problem has as yet only been obtained under very restrictive assumptions in the case of networks. For individual intersections, a more realistic treatment has been possible and is certainly of value for a better understanding of the problem. Whether the solutions are also applicable for practical purposes largely depends on the assumption of known, time-dependent arrival rates and the effects of random variations on operations.

E.3. TRAFFIC-CONTROLLED STRATEGIES

Strategies attempting to minimise delay for an isolated, congested intersection have been suggested by Miller, Weinberg *et al* and Ross *et al*, among others. The references mentioned may be seen as extension of the work by Dunne and Potts for non-saturated conditions. Based on detector output, that control is selected which is predicted to minimize delay for the next phase.

In the Critical Intersection Control (C.I.C.) strategy developed by Ross *et al*, for example, the decision, to be calculated on-line for each time slice of a few seconds, is whether or not to extend the on-going green phase. The number of arrivals during each of a number of future time slices is calculated, for example by using exponential smoothing. Delay is calculated for each of these slices, whereby a number of cases are separated depending on the projected arrival time in relation to the eventual extension and the time to disperse existing queues. Straightforward vehicle trajectory models are used.

Criteria regarding maximum queue length lead to quite different strategies and have been explored by Gordon and Longley for the case of individual intersections. Cycle length c is assumed fixed, and using queue lengths as input the optimum split is calculated cycle by cycle. Gordon's results will be described here as an example. Let x_{ij} be the change in queue length in approach i (= 1,2) from cycle (j-1) to cycle j, Δa_{ij} the difference in number of arrival from cycle (j-1) to cycle j (obtained by an upstream detector), and Δd_{ij} similarly the backward difference in output (obtained by a stop-line detector). Then obviously :

$$x_{ij} = \Delta a_{ij} - \Delta d_{ij} \tag{11}$$

The control variable can be chosen as :

$$Z_1 = g_1/c \tag{12}$$

where

$$g_1 + g_2 = c - L$$
$$\gamma_1 + \gamma_2 = 1 - L/c \tag{13}$$

Gordon chooses as his objective to keep the ratio x_{ij} / x_{2j} between the two queues as near as possible to a known constant K, which would be the ratio between the respective storage spaces. The state variable ("error" in conventional control theory) $_g$ can then be defined through :

$$\Delta r_j = \Delta x_{ij} - k \Delta x_{2g} \tag{14}$$

$$\Delta r_j = r_j - r_{j-1}, \tag{15}$$

If capital letters are used to denote the Z transfer function, i.e.

$$F(Z) = \sum_{k=o}^{\infty} f(kc)\, Z^{-k}$$

it follows from (15) that

$$\frac{R}{\Delta R} = \frac{Z}{Z-1} \tag{16}$$

The problem is now to find a transfer function H (Z) so that

$$\Delta Y = HR \tag{17}$$

where ΔY and R are the transforms of the required change in the control variable y and of the state variable r, respectively.

For a saturated intersection, the number of vehicles serviced during one phase will be :

$$\Delta d_{ij} = g_{i,j-1} \cdot s_i = c\,\gamma_{i,j-1} \cdot s_i \tag{18}$$

where

s_i = saturation flow for approach i.

Neglecting the effect of minimum times, the following transfer function is obtained from (11) - (18) :

$$R(Z) = \frac{[\Delta A_1(Z) - k\Delta A_2(Z) + c\, s_2 (1 - L/c)\, \underline{k}]\, Z\, (Z-1)}{Z^2 + Z\,[(s_1 + k s_2)\, cH(Z) - 2] + 1} \tag{19}$$

It remains to find a suitable H(Z), by considering the characteristic equation (set the denominator of eq. 19 = 0). Several functions would give a stable response. Gordon selects

$$H(Z) = H_o + H_1(Z)$$

where constants H_o and H_1 are chosen to obtain desired frequency and settling time.

The characteristic equation then becomes

$$Z^2 + (BH_o - 2) Z + BH_1 + 1 = 0 \qquad (20)$$

where

$$B = s_1 + ks_2$$

The control will be stable if - and only if - the roots of eq. 20 are within the unit circle, which limits the choice of constants H_o and H_1. Gordon also discusses the effects of green time restrictions on the solution.

Longley uses the same approach but measures the queue lengths at the end of each green period (when they are the smallest), which should have practical advantages. He also extends this technique to a network of four intersections, where vehicles leaving one intersection are assumed to reach the next after a certain number of cycles. All intersections have the same constant cycle time and all offsets are zero. It is shown that the positive feedback produced by the intersection of flows may severely affect stability, but that instability can be reduced through coordination.

Other on-line strategies use more or less heuristic control logics or more restrictive assumptions. David and Henry extend the SATURNE method described above to a network, measuring the queues on each link and calculating green splits so as to minimise total queue lengths during the next cycle. All links are assumed to be saturated.

Again, it would seem that the theoretical development so far has contributed essentially through a better understanding of the problems involved, but not so much in the way of solutions. The most obvious drawback of those on-line models described here, and perhaps also of those empirical methods being implemented, is the inability to utilise sensibly those non-stochastic daily patterns leaned on by the on-line methods. To fill that gap in the middle of the on-off line spectrum would probably require an approach different from any of those described in this Appendix.

APPENDIX F

TRAFFIC CORRIDORS

F.1. OBJECT FUNCTION

Calculation of an object function is shown in Figure F.1 and can be done by a control model in different steps (see Figure F.2) :

- Traffic flow present in the network is constantly recorded by loop detectors placed at relevant cross-sections and traffic data are transmitted to a central process control computer.

- Data are checked and calculated for all motorway sections.

- On the basis of this traffic analysis a short-term prognosis - perhaps for the next five minutes - is accomplished which gives detailed information especially on the imminent density at critical cross-sections (so-called weak points). Relations between k, q and v (see Figure F.3) are given for each section, and volume and speed are available.

- By means of the predicted traffic data it is possible to calculate the objective function data as travel times, consumption of fuel and probability of collisions for different alternatives.

- In the decision phase optimum routes can be elected for traffic guidance or optimum routes can be fixed for access control. Optimal traffic flow is reached for a section when density k_{opt} is reached in the fundamental diagram (see Figure F.3). But greater density leads to oversaturation.

F.2. FREQ MODEL

In California two freeway corridor models have been most extensively developed - FREQ6PL and FREQ6PE. These operational planning models are descendants of a line of models which had been developed previously, starting in 1968. The FREQ6PL model is a macroscopic model of a freeway corridor used primarily for the evaluation of reserved lanes on freeways for carpools and/or buses. The FREQ6PE model is also a macroscopic model of a freeway corridor, but is used primarily for the evaluation of priority entry and normal entry control on a freeway. The input requirements for these models are similar and are described below.

The major data requirement is a set of one to six O-D tables for each interval of time. (15 minutes is recommended as the best time interval.) These tables would correspond to single-occupant cars, two-occupant cars, three-occupant cars, etc., with the sixth or final table referring to buses. Alternatively, the user may input 1 O-D table along with the percentage of single, 2, 3, etc.-occupant autos and the percentage of buses.

Another critical input element is the specification of freeway section capacities. If the user desires the model will also calculate the effect of weaving on capacity. The user may select one of five prespecified speedflow curves or may input his own. Finally, the user must input ramp characteristics. For example, left-hand on-ramps must be noted and ramp capacities must be specified.

For FREQ6PL, the user must specify the priority lane design configuration, the number of people that constitute a legal carpool, and the time duration of priority operations. The model adjusts supply and demand and predicts a time stream of impacts and both spatial and modal traveller responses.

For FREQ6PE, the user must specify the type of entry control and the design changes, if any. Through a linear programming optimisation process, the model will select a ramp control plan. The model also predicts a time stream of impacts and traveller responses.

Figure F.1 **OBJECTIVE FUNCTION**

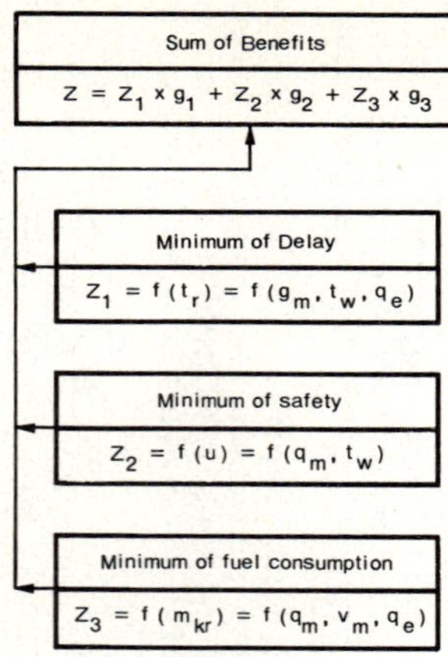

Z, Z_1, Z_2, Z_3 = Benefits ;
g_1, g_2, g_3 = Weights ;
t_r = Travel Time ;
q_m = Volume ;
t_w = Occupancy ;
q_e = Input volume ;
u = Collisions ;
m_{kr} = Consumption of fuel ;
v_m = Speed.

Figure F.2 **CONTROL MODEL**

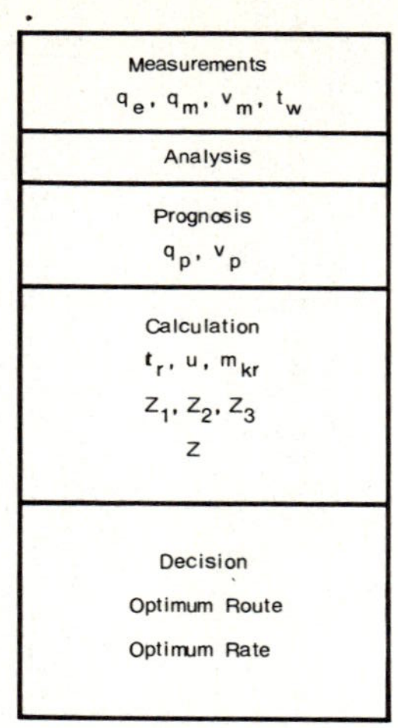

Figure F.3 **FUNDAMENTAL DIAGRAM**

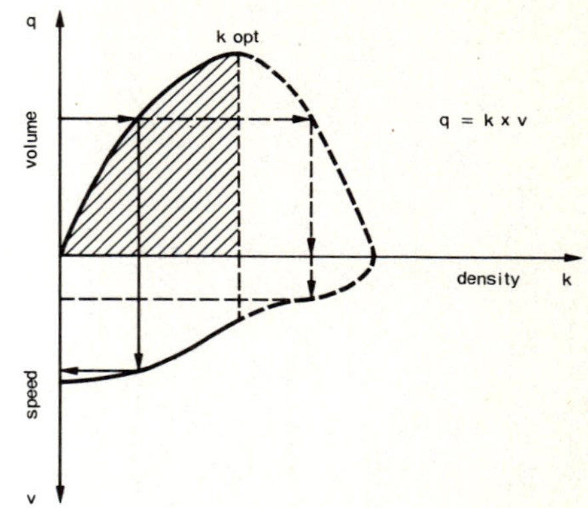

Relation between q, k and v

Based on these general conceptions and models a lot of practical approaches can be described where more or less successfully oversaturation problems have been solved or where attempts to solve them are being made.

As such a study tool the FREQ models tend to become standard as SIGOP and TRANSYT have sometimes been for urban networks.

F.3. BROADCASTING

In France, traffic congestion concentrates on holidays and week-ends. Via newspapers, broadcasting, and television (bison futé), the authorities try to influence road users in advance or during their journey to come to a reasonable decision as to travel routes and times of departure. These measures are intended to lead to a better balanced spatial and temporal distribution of traffic, especially with regard to the desired relief of known bottlenecks.

In Germany, the region-based traffic information identification system for normal broadcasts, ARI (Autofahrer-Rundfunk-Informations-system) has been evolved, followed by others (Austria, Switzerland). With the aid of this area-based information system, road users can be informed in time about general or local critical situations. Furthermore, by these detailed announcements, the road user is enabled to get a general idea of the traffic situation. Thus he can choose respectively to change the route and time of his journey according to circumstances. At the moment, the principal purpose is to improve the accuracy of information by basing it on the most recent situation reports, and to render reliable forecasts possible. ARI essentially makes the accessibility of information easier for the driver.

In the United Kingdom a new method for traffic information broadcasting is in development, the CARFAX system. This system needs its own broadcasting channels.

As EURO-COST30 results can be interpreted the broadcasting solutions vary in different countries, but the objectives are virtually the same.

F.4. VARIABLE MESSAGE SIGNS

In France, information on traffic problems on the road sections ahead is given. This method is a rather linear control procedure. It is, however, also suited for networks. If a bottleneck is indicated in good time, i.e. before the next exit, users can leave the freeway. Such measures have been applied on the north-freeway of Paris and showed good results. On national roads variable message signs are used to indicate alternative routes once congestion appears at certain critical sections.

Starting respectively in 1972 and in 1977, changeable message signs have been operated in the two German project areas of "Rhein-Main" and "Koblenz-Köln". In the Rhein-Main network a short-term prognosis is accomplished on the basis of an overall automatic traffic analysis giving information on the imminent density especially at critical cross-sections. By means of these density data, an on-line calculation of the most favourable routes for users is made. These routes are the basis for the variable traffic guidance of drivers who are already in the network or entering it. On the whole, the system works satisfactorily. At present, an examination of the system's economy of operation is being undertaken. Its range of application is planned to be extended by the regional traffic information broadcasts already described. Attempts will be made to increase its effectiveness in the network itself ; for example, by answering users' information requirements which cannot be satisfied with variable message signs.

EURO-COST30 here aims at the standardisation of a system in order to achieve better utilisation of the road network and to improve traffic safety.

In Japan, changeable message signs are part of the Comprehensive Automobile Traffic Control System CACS. Within the Route Display Board Sub-system the signs recommend routes, and warn of congestion points and accident locations based on traffic information acquired through the Route Guidance Sub-system which calculates optimum routes. Also a number of variable message signs have, since 1967, been put into use on the Tokyo Expressway and there on ramp entrances. To help road users select their best routes, these display location, cause and queue length of congestion ahead. Since at least two out of the seven radial routes of the Tokyo Expressway network function as inter-city routes (Tokyo-Yokohama and Tokyo-Chiba), these signs can be regarded as inter-city devices.

The Integrated Motorist Information System (IMIS) in the United States combines state-of-the-art techniques and equipment to provide cost-effective traffic management and congestion control in a heavily travelled urban corridor. Using automatic vehicle sensors, IMIS will be capable of assessing traffic conditions throughout the corridor and detecting incidents. By means of special changeable message signs, the system will advise motorists of the traffic situation and aid them in selecting the least congested routes for a faster and safer trip. Traffic signals along major arteries in the corridor and metering signals at freeway entrance ramps will also be automatically controlled in response to current traffic demands in order to improve traffic flow further.

The system will be implemented in the Northern Long Island Corridor. This 30 mile long, 5 mile wide corridor is located in a suburban portion of the New York City metropolitan area. Computer simulations, which were employed to evaluate achievable benefits of IMIS under various traffic conditions at various times of the day in the corridor, produced the following estimates of annual benefits which could be obtained :

- 2.9 million vehicle hours of delay saved
- 200 fewer accidents
- 2.7 million gallons of fuel saved
- 6 million pounds of pollutants eliminated

The IMIS control logic consists of two levels of control. The corridor level allocates traffic among the various corridor facilities. The local level optimises the usage of individual facilities independently, based on predicted demand as determined at the corridor level. The system is thus hierarchical. Attempts are made to re-allocate traffic every 10-20 minutes or whenever measured conditions warrant change. Corridor-level control consists of demand estimation, corridor control, and corridor control message selection ; local-level control consists of ramp metering control, arterial signalisation control, incident detection, and state estimation.

Demand estimation refers to the prediction of O-D patterns as a function of time. The predictions are based on a combination of historical data and synthesised data derived from estimates of on-and-off flows at system entrance and exits and estimates of link volumes.

The objective of the corridor control function is to determine the amount of diversion required at the major potential diversion nodes in the network in order to optimise the distribution of traffic flow in the network.

F.5. INDIVIDUAL GUIDANCE

In the German Ruhrgebiet project area the prior conditions for an individual destination guiding system for highway traffic are being established. Vehicles equipped with special devices are enabled to feed their destination intentions into the system. After that they will receive information concerning the conditions on the road sections ahead. At interchanges they will get recommendations concerning the next sections of route to be followed. This information is provided on special display units in the vehicles.

The determination of the most favourable route takes place according to the control model developed in the Rhein-Main network. The fact that information about conditions and guiding recommendations are offered individually to the users concerned is regarded as the system's main advantage because of its anticipated high rate of acceptance. Furthermore, the input of destination intentions leads to the assumption that improved O-D-input data for the present total traffic can be expected.

In Japan a similar system was also part of CACS. The Driving Information Sub-system feeds information, for safe driving and the optimum route, to a display inside the vehicle. The field experiment was carried out in Tokyo for a year from Oct. 1977 covering an area of 30 km^2 with 85 route-guiding intersections and expressway forks and with 1,300 test vehicles which were equipped with on-board devices. Reliability of the hardware system as well as time-saving benefit to users was found quantitatively through the experiment. The MTBF of the roadside equipment was found to be 24,000 hours and that of the on-vehicle equipment was estimated to be 2 years on the assumption that the average usage is 1.5 hours a day. The average travel time-saving was 13 percent compared with non-guided vehicles in daytime conditions.

F.6. ACCESS CONTROL

In France, access control is related to freeways and the secondary parallel network. In order to maintain a constant traffic flow the admission at access roads will be controlled. Furthermore, in case of congestion, vehicles are guided into the parallel network at exits. The procedure is similar to the methods applied in urban networks.

In May and June 1978, a corresponding experiment was carried through on the main freeways of a Paris suburb. Access control was connected with traffic management and traffic control operations and supported by measures to improve traffic conditions on parallel roads. It is not yet possible to regard the result of this special experiment as generally valid. However, it must be noted that travel times on the freeways decreased considerably without at the same time increasing on the parallel roads. User opinion was generally favourable. Similar experiments, especially on the Paris north-freeway, also led to positive results.

In France, also, the PERCEVAL program developed by IRT should be mentioned. This program carries through a calculation of admission rates at different approach roads on the basis of a given O-D-matrix.

In Germany, access control measures at the preceding access road were applied in 1968/69 at a temporarily much overloaded freeway intersection (Kamener Kreuz). It was possible to prevent traffic jams on the critical freeway section and,significantly, to increase the capacity of the bottleneck. The results, however, can only, to a certain extent, be considered as generally relevant. At the time of the experiment, problems of traffic control in the secondary network concerned, as well as the economic aspects of such operations, have not been included in the plans for further investigation.

In Canada Q.E.W. Freeway Surveillance and the Traffic Control Demonstration Project became operational in July 1975 after the benefits of other control systems, installed in Chicago, Dallas, Detroit, Houston, Los Angeles and New York had been demonstrated.

Here an acceptable level of freeway service was maintained by ramp metering. The traffic control strategy is based on a traffic response plan selection in which ramp signal cycle lengths are selected on the basis of both mainline and on-ramp traffic conditions.

From the results it is apparent that, even under poor conditions, a significant improvement in overall performance has been achieved by ramp metering. Even ramp delays had decreased after installation of the system. In view of these favourable results additional equipment will be added and a local expansion program will be undertaken.

In California, the FREQ6PL model has been applied to the Santa Monica Freeway in Los Angeles and, through sensitivity analysis, policy guidelines have been proposed.

The FREQ6PE model has been applied to the Eastshore Freeway in the San Francisco Bay Area and here also some initial policy guidelines have been proposed.

The IMIS ramp metering control function calculates metering rates for the metered ramps along a freeway. The rates are based on the predicted ramp demands and on the predicted freeway demand, accounting for the additional demand expected from upstream traffic diversions.

LIST OF MEMBERS OF THE GROUP

Chairman : Mr. J.J. Klijnhout, The Netherlands

AUSTRIA	Prof. H. Knoflacher Kuratorium für Verkehrssicherheit Olzeltgasse 3 A-1031 Vienne
BELGIUM	Mr. M. Fontaine Service du Trafic Routier Ministère des Travaux Publics Chaussée de Louvain 550 B-1030 Bruxelles
	Mr. P. Tielemans Ingénieur en Chef-Directeur des Ponts et Chaussées Service du Trafic Routier Ministère des Travaux Publics Chaussée de Louvain 550 B-1030 Bruxelles
CANADA	Mr. J.A. Sharp Head Transportation Planning Technology Mohawk College 135 Fennell Avenue, W. Hamilton Ontario L8N 3T2
FRANCE	M. Laurent Amy de la Breteque Département Métrologie et Automatisme I.R.T. B.P. 28 Avenue du Général Malleret-Joinville 94114 Arcueil Cedex
	M. Texier CETUR 8 avenue Aristide Briand 92220 Bagneux
GERMANY	Dipl. Ing. J. Boesefeldt Peterstrasse 2/4 51 Aachen
IRELAND	Mr. M. Staunton Research Officer Road Traffic Section An Foras Forbartha St. Martin's House Waterloo Road Dublin 4
ITALY	Ing. Franco Treglia Società Autostrade Via Nibby 10 I-00161 Rome

JAPAN	Mr. Shunji Hashiguchi Associate Director Traffic Control Division Traffic Bureau National Police Agency 2-1-2 Kasumigaseki Chiyoda-ku Tokyo 100
	Mr. Koichi Tsuchiya Deputy Head Road Traffic Control Division Road Bureau Ministry of Construction 2-1-3 Kasumigaseki Chiyoda-Ku Tokyo 100
NETHERLANDS	Ir. J.J. Klijnhout Dienst Verkeerskunde van de Rijkswaterstaat Postbus 20906 2500 EX 's-Gravenhage
SPAIN	M. J. Garcia Ayuntamiento de Barcelona Avenida Puerta del Angel N° 8-10 - 4° Barcelona
	D. Juan G. Gardeta Civil Engineer of the Traffic Section Laboratorio del Transporte y Mecanica del Suelo "Jose Luis Escario" Centre de Estudios y Experimentacion de Obras Publicas Alfonso XII 3 Madrid 7
	M. José Luis Pedragosa Sub-Director-General Dirección General de Tráfico Ministry of the Interior Amador de los Rios 1 Madrid
	M. Sebastian de la Rica Jefe de la Seccion de Circulacion Delegacion de Servicios de Circulacion y Transporte Excmo. Ayuntamiento de Madrid Plaza de la Villa 1 Madrid
SWEDEN	Mr. Arne Hansson AB Vattenbyggnadsbyran Geijersgatan 8 S-216 18 Malmo
UNITED KINGDOM	Dr. P. Hunt Urban Network Division Transport and Road Research Laboratory (TRRL) Old Wokingham Road Crowthorne Berks. RG11 6AU
UNITED STATES	Mr. Gary Euler HRS-31 Traffic Systems Division Federal Highway Administration U.S. Department of Transportation Washington D.C. 20591

UNITED STATES (Cont'd)	Mr. C.J. MacGowan Highway Research Engineer Federal Highway Administration U.S. Department of Transportation Washington D.C. 20591
	Mr. W.W. Wolman Federal Highway Administration U.S. Department of Transportation Washington D.C. 20591
OECD	Mr. B. Horn
Rapporteurs of the report were :	Messrs. Klijnhout, Horn, Texier, Gardeta, Garcia, Pedragosa, de la Rica, Staunton, Sharp, MacGowan, Knoflacher, Boesefeldt, Hansson, Amy de la Breteque, Tielemans, Fontaine, Hunt.
Members of the Editing Committee were :	Messrs. Klijnhout, Euler, Sharp, Hunt; Amy de la Breteque and Texier (for the French version); and Horn.

LIST OF ROAD RESEARCH PUBLICATIONS

Road Construction and Maintenance

Research on Crash Barriers (February 1969)
Motor Vehicle Corrosion and Influence of De-icing Chemicals (October 1969)
Winter Damage to Road Pavements (May 1972)
Accelerated Methods of Life-testing Pavements (May 1972)
Proceedings of the Symposium on the Quality Control of Road Works (July 1972)
Waterproofing of Concrete Bridge Decks (July 1972)
Optimisation of Road Alignment by the Use of Computers (July 1973)
Water in Roads : Prediction of Moisture Content in Road Subgrades (August 1973)
Maintenance of Rural Roads (August 1973)
Water in Roads : Methods for Determining Soil Moisture Content and Pore Water Tension (December 1973)
Proceedings of the Symposium on Frost Action on Roads (October 1974)
Road Marking and Delineation (February 1975)
Resistance of Flexible Pavements to Plastic Deformation (June 1975)
Bridge Inspection (July 1976)
Road Strengthening (September 1976)
Proceedings of the Symposium on Road Design Standards (May 1977)
Use of Waste Materials and By-products in Road Construction (September 1977)
Maintenance Techniques for Road Surfacings (October 1978)
Catalogue of Road Surface Deficiences (October 1978)
Evaluation of Load Carrying Capacity of Bridges (December 1979)
Construction of Roads on Compressible Soils (December 1979)

Road Safety

Alcohol and Drugs (January 1968)
Pedestrian Safety (October 1969)
Driver Behaviour (June 1970)
Proceedings of the Symposium on the Use of Statistical Methods in the Analysis of Road Accidents (September 1970)
Lighting, Visibility and Accidents (March 1971)
Research into Road Safety at Junctions in Urban Areas (October 1971)
Road Safety Campaigns : Design and Evaluation (December 1971)
Speed Limits Outside Built-up Areas (August 1972)
Research on Traffic Law Enforcement (April 1974)
Young Driver Accidents (March 1975)
Roadside Obstacles (August 1975)
Manual on Road Safety Campaigns (September 1975)
Polarised Light for Vehicle Headlamps (December 1975)
Driver Instruction (March 1976)
Adverse Weather, Reduced Visibility and Road Safety (August 1976)
Hazardous Road Locations : Identification and Countermeasures (September 1976)
Safety of Two-Wheelers (March 1978)
New Research on the Role of Alcohol and Drugs in Road Accidents (September 1978)
Traffic Safety in Residential Areas (October 1979)
Road Safety at night (December 1979)

Road Traffic and Urban Transport

Electronic Aids for Freeway Operation (April 1971)
Area Traffic Control Systems (February 1972)
Optimisation of Bus Operation in Urban Areas (May 1972)
Two Lane Rural Roads : Road Design and Traffic Flow (July 1972)
Traffic Operation at Sites of Temporary Obstruction (February 1973)
Effects of Traffic and Roads on the Environment in Urban Areas (July 1973)
Proceedings of the Symposium on Techniques of Improving Urban Conditions by Restraint of Road Traffic (September 1973)
Urban Traffic Models : Possibilities for Simplification (August 1974)
Capacity of At-grade Junctions (November 1974)

Proceedings of the Symposium on Road and the Urban Environment (October 1975)
Research on Traffic Corridor Control (November 1975)
Bus Lanes and Busway Systems (December 1976)
Transport Requirements for Urban Communities (December 1977)
Energy Problems and Urban and Suburban Transport (December 1977)
Integrated Urban Traffic Management (December 1977)
Traffic Measurement Methods for Urban and Suburban Areas (March 1979)
Transport Services in Low Density Areas (September 1979)
Urban Public Transport : Evaluation of Performance (October 1980)

OECD SALES AGENTS
DÉPOSITAIRES DES PUBLICATIONS DE L'OCDE

ARGENTINA – ARGENTINE
Carlos Hirsch S.R.L., Florida 165, 4° Piso (Galería Guemes)
1333 BUENOS AIRES, Tel. 33.1787.2391 y 30.7122

AUSTRALIA – AUSTRALIE
Australia and New Zealand Book Company Pty, Ltd.,
10 Aquatic Drive, Frenchs Forest, N.S.W. 2086
P.O. Box 459, BROOKVALE, N.S.W. 2100

AUSTRIA – AUTRICHE
OECD Publications and Information Center
4 Simrockstrasse 5300 BONN. Tel. (0228) 21.60.45
Local Agent/Agent local :
Gerold and Co., Graben 31, WIEN 1. Tel. 52.22.35

BELGIUM – BELGIQUE
LCLS
35, avenue de Stalingrad, 1000 BRUXELLES. Tel. 02.512.89.74

BRAZIL – BRÉSIL
Mestre Jou S.A., Rua Guaipa 518,
Caixa Postal 24090, 05089 SAO PAULO 10. Tel. 261.1920
Rua Senador Dantas 19 s/205-6, RIO DE JANEIRO GB.
Tel. 232.07.32

CANADA
Renouf Publishing Company Limited,
2182 St. Catherine Street West,
MONTRÉAL, Quebec H3H 1M7. Tel. (514)937.3519
522 West Hasting,
VANCOUVER, B.C. V6B 1L6. Tel. (604) 687.3320

DENMARK – DANEMARK
Munksgaard Export and Subscription Service
35, Nørre Søgade
DK 1370 KØBENHAVN K. Tel. +45.1.12.85.70

FINLAND – FINLANDE
Akateeminen Kirjakauppa
Keskuskatu 1, 00100 HELSINKI 10. Tel. 65.11.22

FRANCE
Bureau des Publications de l'OCDE,
2 rue André-Pascal, 75775 PARIS CEDEX 16. Tel. (1) 524.81.67
Principal correspondant :
13602 AIX-EN-PROVENCE : Librairie de l'Université.
Tel. 26.18.08

GERMANY – ALLEMAGNE
OECD Publications and Information Center
4 Simrockstrasse 5300 BONN Tel. (0228) 21.60.45

GREECE – GRÈCE
Librairie Kauffmann, 28 rue du Stade,
ATHÈNES 132. Tel. 322.21.60

HONG-KONG
Government Information Services,
Sales and Publications Office, Baskerville House, 2nd floor,
13 Duddell Street, Central. Tel. 5.214375

ICELAND – ISLANDE
Snaebjörn Jönsson and Co., h.f.,
Hafnarstraeti 4 and 9, P.O.B. 1131, REYKJAVIK.
Tel. 13133/14281/11936

INDIA – INDE
Oxford Book and Stationery Co. :
NEW DELHI, Scindia House. Tel. 45896
CALCUTTA, 17 Park Street. Tel. 240832

INDONESIA – INDONÉSIE
PDIN-LIPI, P.O. Box 3065/JKT., JAKARTA, Tel. 583467

IRELAND – IRLANDE
TDC Publishers – Library Suppliers
12 North Frederick Street, DUBLIN 1 Tel. 744835-749677

ITALY – ITALIE
Libreria Commissionaria Sansoni :
Via Lamarmora 45, 50121 FIRENZE. Tel. 579751
Via Bartolini 29, 20155 MILANO. Tel. 365083
Sub-depositari :
Editrice e Libreria Herder,
Piazza Montecitorio 120, 00 186 ROMA. Tel. 6794628
Libreria Hoepli, Via Hoepli 5, 20121 MILANO. Tel. 865446
Libreria Lattes, Via Garibaldi 3, 10122 TORINO. Tel. 519274
La diffusione delle edizioni OCSE è inoltre assicurata dalle migliori librerie nelle città più importanti.

JAPAN – JAPON
OECD Publications and Information Center,
Landic Akasaka Bldg., 2-3-4 Akasaka,
Minato-ku, TOKYO 107 Tel. 586.2016

KOREA – CORÉE
Pan Korea Book Corporation,
P.O. Box n° 101 Kwangwhamun, SÉOUL. Tel. 72.7369

LEBANON – LIBAN
Documenta Scientifica/Redico,
Edison Building, Bliss Street, P.O. Box 5641, BEIRUT.
Tel. 354429 – 344425

MALAYSIA – MALAISIE
and/et SINGAPORE - SINGAPOUR
University of Malaysia Co-operative Bookshop Ltd.
P.O. Box 1127, Jalan Pantai Baru
KUALA LUMPUR. Tel. 51425, 54058, 54361

THE NETHERLANDS – PAYS-BAS
Staatsuitgeverij
Verzendboekhandel Chr. Plantijnnstraat
S-GRAVENAGE. Tel. nr. 070.789911
Voor bestellingen: Tel. 070.789208

NEW ZEALAND – NOUVELLE-ZÉLANDE
Publications Section,
Government Printing Office,
WELLINGTON: Walter Street. Tel. 847.679
Mulgrave Street, Private Bag. Tel. 737.320
World Trade Building, Cubacade, Cuba Street. Tel. 849.572
AUCKLAND: Hannaford Burton Building,
Rutland Street, Private Bag. Tel. 32.919
CHRISTCHURCH: 159 Hereford Street, Private Bag. Tel. 797.142
HAMILTON: Alexandra Street, P.O. Box 857. Tel. 80.103
DUNEDIN: T & G Building, Princes Street, P.O. Box 1104.
Tel. 778.294

NORWAY – NORVÈGE
J.G. TANUM A/S Karl Johansgate 43
P.O. Box 1177 Sentrum OSLO 1. Tel. (02) 80.12.60

PAKISTAN
Mirza Book Agency, 65 Shahrah Quaid-E-Azam, LAHORE 3.
Tel. 66839

PHILIPPINES
National Book Store, Inc.
Library Services Division, P.O. Box 1934, MANILA.
Tel. Nos. 49.43.06 to 09, 40.53.45, 49.45.12

PORTUGAL
Livraria Portugal, Rua do Carmo 70-74,
1117 LISBOA CODEX. Tel. 360582/3

SPAIN – ESPAGNE
Mundi-Prensa Libros, S.A.
Castello 37, Apartado 1223, MADRID-1. Tel. 275.46.55
Libreria Bastinos, Pelayo 52, BARCELONA 1. Tel. 222.06.00

SWEDEN – SUÈDE
AB CE Fritzes Kungl Hovbokhandel,
Box 16 356, S 103 27 STH, Regeringsgatan 12,
DS STOCKHOLM. Tel. 08/23.89.00

SWITZERLAND – SUISSE
OECD Publications and Information Center
4 Simrockstrasse 5300 BONN. Tel. (0228) 21.60.45
Local Agents/Agents locaux
Librairie Payot, 6 rue Grenus, 1211 GENÈVE 11. Tel. 022.31.89.50
Freihofer A.G., Weinbergstr. 109, CH-8006 ZÜRICH.
Tel. 01.3624282

TAIWAN – FORMOSE
National Book Company,
84-5 Sing Sung South Rd, Sec. 3, TAIPEI 107. Tel. 321.0698

THAILAND – THAILANDE
Suksit Siam Co., Ltd., 1715 Rama IV Rd,
Samyan, BANGKOK 5. Tel. 2511630

UNITED KINGDOM – ROYAUME-UNI
H.M. Stationery Office, P.O.B. 569,
LONDON SE1 9NH. Tel. 01.928.6977, Ext. 410 or
49 High Holborn, LONDON WC1V 6 HB (personal callers)
Branches at: EDINBURGH, BIRMINGHAM, BRISTOL, MANCHESTER, CARDIFF, BELFAST.

UNITED STATES OF AMERICA – ÉTATS-UNIS
OECD Publications and Information Center, Suite 1207,
1750 Pennsylvania Ave., N.W. WASHINGTON D.C.20006.
Tel. (202) 724.1857

VENEZUELA
Libreria del Este, Avda. F. Miranda 52, Edificio Galipan,
CARACAS 106. Tel. 32.23.01/33.26.04/33.24.73

YUGOSLAVIA – YOUGOSLAVIE
Jugoslovenska Knjiga, Terazije 27, P.O.B. 36, BEOGRAD.
Tel. 621.992

Les commandes provenant de pays où l'OCDE n'a pas encore désigné de dépositaire peuvent être adressées à :
OCDE, Bureau des Publications, 2, rue André-Pascal, 75775 PARIS CEDEX 16.

Orders and inquiries from countries where sales agents have not yet been appointed may be sent to:
OECD, Publications Office, 2 rue André-Pascal, 75775 PARIS CEDEX 16.

OECD PUBLICATIONS, 2 rue André-Pascal, 75775 PARIS CEDEX 16 - No. 41 761 1981
PRINTED IN FRANCE
(900 TS 77 81 02 1) ISBN 92-64-12154-4